THE UNIVERSE OF GALAXIES

Readings from
**SCIENTIFIC
AMERICAN**

THE UNIVERSE OF
GALAXIES

Compiled by
Paul W. Hodge
University of Washington

W. H. Freeman and Company
New York

Some of the SCIENTIFIC AMERICAN articles in
The Universe of Galaxies are available as separate
Offprints. For a complete list of articles now available as
Offprints, write to W. H. Freeman and Company, 41
Madison Avenue, New York, New York 10010.

Library of Congress Cataloging in Publication Data

Main entry under title:

The Universe of galaxies.

   "Readings from Scientific American."
   Bibliography: p.
   Includes index.
   1. Galaxies.   I. Hodge, Paul W. II. Scientific
American.
QB857.U55 1984   523.1'12   84-4090
ISBN 0-7167-1675-5
ISBN 0-7167-1676-3 (pbk.)

Printed in the United States of America.

CONTENTS

Note on cross-references to SCIENTIFIC AMERICAN *articles*: Articles included in this book are referred to by title and page number; articles not included in this book but available as Offprints are referred to by title and offprint number; articles not included in this book and not available as Offprints are referred to by title and date of publication.

PREFACE

The well-known persuasive power of the editors of SCIENTIFIC AMERICAN has led to the recent publication in that magazine of a number of superb articles on galaxies, written by some of the most active and imaginative scientists in the field. This book is a collection drawn from these articles and mounted in a sequence that stretches from our own galaxy to the far reaches of the cosmos. It is intended to provide a picture not only of the universe, but also of the ways in which we are learning about it, as told by the modern-day pioneers who are engaged in exploring it.

I am indebted to the staff at W. H. Freeman and Company for inviting me to put this book together, to the editors of SCIENTIFIC AMERICAN for seeking such authoritative writers and to the authors themselves for presenting current research on galaxies in a vital and engrossing manner. The difficult job was in limiting the volume to a mere nine articles; many others that have appeared in the magazine would have added a great deal to the collection. I urge the reader, once he or she has finished reading this book, to explore other issues of SCIENTIFIC AMERICAN (including ones antedating those excerpted here) for additional fine reading.

Paul W. Hodge
Professor of Astronomy
University of Washington

THE UNIVERSE OF GALAXIES

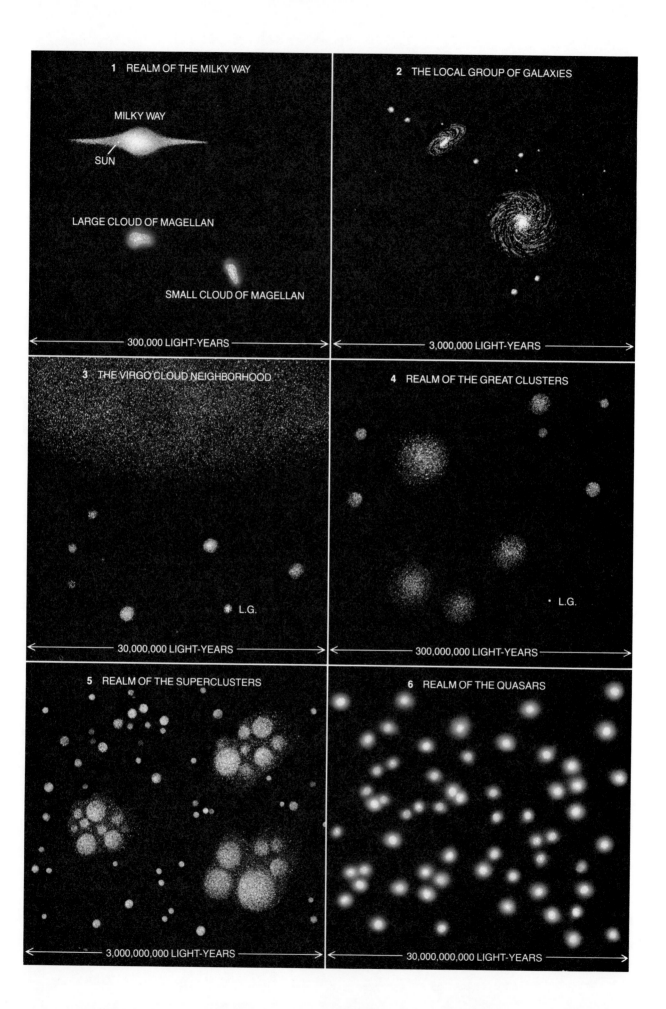

1   REALM OF THE MILKY WAY

MILKY WAY

SUN

LARGE CLOUD OF MAGELLAN

SMALL CLOUD OF MAGELLAN

← 300,000 LIGHT-YEARS →

2   THE LOCAL GROUP OF GALAXIES

← 3,000,000 LIGHT-YEARS →

3   THE VIRGO CLOUD NEIGHBORHOOD

L.G.

← 30,000,000 LIGHT-YEARS →

4   REALM OF THE GREAT CLUSTERS

L.G.

← 300,000,000 LIGHT-YEARS →

5   REALM OF THE SUPERCLUSTERS

← 3,000,000,000 LIGHT-YEARS →

6   REALM OF THE QUASARS

← 30,000,000,000 LIGHT-YEARS →

INTRODUCTION

I t was not until this century that galaxies were recognized as the chief form of organized matter in the universe. But in the years since their discovery, galaxies have become among the most vigorously studied astronomical entities. For several decades their faintness and great distance made them accessible only to a few large optical telescopes and a small number of privileged astronomers. Since about 1960, however, new types of large telescopes, including radio and space-borne instruments, have tremendously increased the opportunities to understand the galaxies. As a result, a large number of astronomers are now devoting themselves to this enterprise.

To illustrate the dramatic intensification in the study of galaxies, I refer you to the fact that *The Astrophysical Journal*, the premier American journal of professional astronomy, carried only one article on galaxies in all of 1930. This paper had just one author—E. P. Hubble of the Mount Wilson Observatory in southern California. Fifty years later, in 1980, the same journal published 276 articles dealing with galaxies; these were authored by 423 astronomers from institutions all over the world. Of course, the above figures reflect in part the stupendous growth in all fields of astronomy: in 1930 the *Journal* had 704 pages, whereas in 1980 its page count reached nearly 15,000. Nevertheless, the increase in galaxy research, as measured in terms of journal space assigned to it, has been about 10 times as great as the increase in astronomical research overall.

Most of our knowledge of the properties of galaxies is of very recent origin. It should not be surprising that this collection is made up primarily of SCIENTIFIC AMERICAN articles published within the last 10 years. In fact, all but two of them were written in the 1980's. Other interesting and relevant articles that have appeared in the magazine include: "The Dynamics of the Andromeda Nebula" (Rubin; June, 1973), "Giant Radio Galaxies" (Strom, Miley and Oort; August, 1975), "BL Lacertae Objects" (Disney and Veron; August, 1977), "The Clustering of Galaxies" (Groth, Peebles, Seldner and Soneira; November, 1977) and "Coronas of Galaxies" (de Boer and Savage; August, 1982).

The nine articles in this book explain most of the exciting new developments in the astronomy of galaxies and provide good coverage of the fundamental properties of galaxies. The first four articles present basic facts

DIFFERENT-SIZED REALMS OF THE UNIVERSE OF GALAXIES are shown schematically. Each panel is 10 times larger than the preceding one. The local group of galaxies (L. G.) is a small cluster near the outskirts of a loose aggregate of clusters centered on the Virgo Cloud. There is not yet agreement on its size, nor on the sizes of many clusters and superclusters. It is possible that superclusters may be more netlike, with filaments surrounded by voids.

about normal galaxies—their structure, their content and the mysterious "missing mass"—and set forth ideas about their evolution. Comparing the first and second articles will give a reader an impression of the similarities between astronomers' problems in trying to unravel the structure of our own Galaxy and in interpreting the spiral arms of a neighboring galaxy. Such a comparison will also illustrate some of the differences in perspective that characterize the two tasks: in the first case we are surveying the forest from a position buried deeply among the trees, whereas in the second case we are viewing the forest from afar.

One of the most puzzling and frustrating properties of normal galaxies is described in the third article. Only recently has it been determined that most of a galaxy is invisible. Apparently, and surprisingly, a huge outer "halo" that emits no light surrounds the visible disk of a spiral galaxy like ours. This halo contains most of the mass of the galaxy. The phenomenon is puzzling because no one has yet come up with a convincing or even a plausible idea of what exotic material makes up this mass (and, thus, makes up most of the matter in the universe).

A final article (the fourth) on normal galaxies provides some recent results on a most important and difficult topic, the evolution of galaxies. (Our ideas have evolved much faster than their subject; it is fascinating to compare this article with two previous SCIENTIFIC AMERICAN essays: "The Evolution of Galaxies" by J. H. Oort in September, 1956, and "The Evolution of Galaxies" by H. C. Arp in January, 1963.) New concepts introduced here include the changes in the structure and appearance of galaxies brought on by galactic winds, spiral waves and the ram pressure of intergalactic gas. Although the evolution of galaxies is still a field in its infancy, we have increasing evidence that the development of many galaxies depends as much on "environment" as on "genetics."

The most dynamic examples of this point are beautifully displayed in the fifth article. When two galaxies stray too close to each other, the "evolution" of both is dramatically affected by gravitational interactions. Many of the weird and baffling galaxies that don't fit into the normal scheme of forms appear to be explained simply and elegantly by computer simulations of catastrophic galaxy encounters. Curious readers will note that M51, the Whirlpool Nebula, is presented as a prime example of such tidal distortion; and that it also appears as an evolutionary example in the preceding chapter.

Normal galaxies, as typified by the Milky Way and the Andromeda galaxies, may not always be "normal." A small but important fraction of them appear to be passing through a phase of violent activity, with the center of action lying somewhere in their nuclei. The nearest example of such an immense, violent galactic event is a bright object in the constellation Centaurus, one of the first and brightest radio sources discovered. This object, now called Centaurus A, was catalogued a century ago as NGC 5128, and classified in the interim as a "peculiar" galaxy. It provides a spectacular, well-studied example of a giant radio galaxy to illustrate the amazing properties of active galaxies, which probably owe their activity to a massive black hole in their centers (article 6).

More distant radio galaxies have many similarities to Centaurus A. The seventh article covers a variety of examples, emphasizing the remarkable physics involved in explaining the immense, focused streams of ionized gas that often protrude from the nuclei of radio galaxies. Included is a particularly well illustrated explanation of the "superluminal expansion" found in cases where, because of projection effects, material appears to emerge from the nuclei at velocities greater than the velocity of light.

Galaxies are not arranged randomly but tend to exist in groups and clusters. Recently the study of these aggregates has turned up a striking and rather surprising result. This new knowledge results largely from the introduction of highly efficient electronic detectors that have made it possible to measure velocities and brightnesses of large samples of galaxies spread out

GIANT AND REMARKABLE 30 DORADUS NEBULA (the "Tarantula nebula") was photographed by the four-meter telescope of the Cerro Tololo Inter-American Observatory in Chile. Recent studies using ground-based and orbiting telescopes have shown that the nebula is powered largely by one or more extremely massive stars. One study has suggested that the central star may have 2,000 times the mass of the sun. The nebula is the largest cloud of glowing gas in the nearby galaxy known as the Large Cloud of Magellan, which is about 150,000 light-years from Earth.

over vast reaches of space. These surveys show that the clusters of galaxies are themselves clustered in even larger groups called superclusters. Of equally staggering dimensions are the recently recognized voids between clusters (article 8).

Since their discovery, the enigmatic quasars (article 9) were uncertain citizens of the cosmos. Most astronomers believed that their enormous red shifts should be interpreted in the standard way, to indicate huge velocities of recession in the expanding universe. However, the immense distances that were inferred from the greatest red shifts—which have values as large as $z = 3.5$ (where z is the change in wavelength divided by the wavelength)—were among the considerations that prompted some astronomers to question the premises underlying the interpretation of cosmic red shifts. They maintained that some new, unexpected and unexplained physics was at work, that the quasars are really much closer than their red shifts indicate and that they are possibly associated with "new matter" recently ejected from nearby galaxies. The article is somewhat cautious about the standard interpretation of quasar distances, in deference to this unconventional hypothesis. However, in the months since the article was written, some telling bits of evidence have emerged to settle the question. Several groups of researchers, using the Canada-France-Hawaii 3.6-meter telescope, the Kitt Peak four-meter telescope and the giant Multi-Mirror Telescope on Mount Hopkins, have located the images of the host galaxies of quasars. It now seems certain that the quasar phenomenon involves a supermassive object at the center of galaxies, that the galaxies are at the distances indicated by their quasar's red shifts and that these violent and brilliant things are the most distant visible objects at the edge of the universe of galaxies.

The Milky Way Galaxy

by Bart J. Bok
March, 1981

*A few years ago the fundamental facts about it seemed
to be fairly well established. Now even its mass and
its radius have come into question*

On a clear and moonless night free of the lights of civilization the most arresting thing in the sky is the luminous band of the Milky Way. Even without a telescope there is much one can say about it. One can see, for example, that stars become more numerous as one directs one's gaze across the sky and into the band. The fact that the band itself is made up mostly of stars then seems less astonishing. One can see that the band follows a great circle that bisects the celestial sphere. Hence the earth is embedded in the central plane of the band. One can see, moreover, that the band is widest and brightest in the direction of the constellation Sagittarius. Surely this is the direction to the center of the system. The system is the Milky Way galaxy.

A telescope will reveal that certain starlike points in the Milky Way are actually great aggregations of stars. They are plainly distant objects; they are known as globular clusters. Between 1918 and 1921 Harlow Shapley employed the telescopes of the Mount Wilson Observatory to demonstrate that the clusters, like the stars in the band, are commonest in Sagittarius. In an area of Sagittarius that makes up only 2 percent of the sky Shapley plotted a third of all the globular clusters then known. Evidently, therefore, the solar system is far from the center of the galaxy; the distance from the sun to the center is now estimated to be 8,500 parsecs. (A parsec is 3.26 light-years.) It has since been established that the solar system, along with the rest of the mass in the central plane of the galaxy, revolves about the center. The sun revolves at a rate now taken to be some 230 kilometers per second. It thus completes a revolution every 200 million years.

In 1930 Robert J. Trumpler of the Lick Observatory showed that an interstellar medium of gas and dust dims the light of the stars, particularly the stars in the central plane of the galaxy. In 1951 William W. Morgan of the Yerkes Ob-

servatory and his students Donald E. Osterbrock and Stewart L. Sharpless found evidence that the band is an edge-on view from the earth of the galaxy's spiral structure. Evidence for spiral features soon followed at radio wavelengths. In the 1960's Chia Chiao Lin of the Massachusetts Institute of Technology and Frank H. Shu, then at the Harvard College Observatory, suggested that waves of increased density in the interstellar medium precipitate the spiral features and are responsible for the formation of stars. Meanwhile the advent of infrared astronomy provided a means for exploring the interior of dark interstellar clouds. It now appears that the clouds are where most of the galaxy's new stars form. By 1970 radio astronomy had begun to reveal the composition of the dark clouds. They are made up of dust and hydrogen, with an admixture of some surprisingly complex molecules, many of them organic.

For many years I have been a night watchman of the Milky Way galaxy. I remember the mid-1970's as a time when I and my fellow watchers were notably self-assured. The broad outlines of the galaxy seemed reasonably well established. The galaxy had two main components: a dense central bulge of stars with a boundary between 4,000 and 5,000 parsecs from the center, and a flat, much thinner disk of stars and interstellar gas and dust whose inner margin abutted the bulge and whose outer margin lay some 15,000 parsecs from the center. The bulge is in Sagittarius; the disk is flung across the sky.

The combined mass of the disk and the bulge was then calculated to be well under 200 billion times the mass of the sun. The disk and the bulge were surrounded, however, by a "halo" of matter on each side of the galaxy's central plane. In overall form the halo is a slightly flattened sphere. In the plane of the galaxy it has a radius on the order of 20,000 parsecs. It is notable for old stars, and also for a scattering of per-

haps 100 globular clusters. (Another 100 globular clusters lie in or near the galactic disk.) The halo might add at most 100 billion solar masses to the total mass of the galaxy.

Much work was still to be done. We felt confident nonetheless that the basic findings would stand, and that astronomers would be undistracted as they looked into such matters as how stars form and how the Milky Way's spiral features evolve. We did not suspect it would soon be necessary to revise the radius of the Milky Way upward by a factor of three or more and to increase its mass by as much as a factor of 10. The revisions are emblematic of a number of recent upheavals. Here I shall take up several aspects of the current effort to understand the Milky Way. I include most of them because they are areas of notable ferment where progress in understanding may be imminent. I include one, in contrast, because I fear it is reaching a dead end.

Galactic Coronas

Suggestions that the Milky Way galaxy is unexpectedly large and massive were offered as early as 1974. On the theoretical side they came principally from Donald Lynden-Bell of the University of Cambridge and from Jeremiah P. Ostriker, P. J. E. Peebles and Amos Yahil at Princeton University. The basic argument was that the dynamical stability and permanence of the galaxy cannot be guaranteed unless the galactic disk is surrounded, and thereby stabilized gravitationally in spite of its thinness and its delicate spiral structure, by an extended and massive halo.

J. Einasto and his associates at the Tartu Observatory in Estonia came forward with a different chain of reasoning. For several years investigators looking into the dynamics of the Milky Way had argued that the velocity of the sun with respect to the globular clusters in the halo of the galaxy is only some 180 kilometers per second. Since the clusters are scattered throughout a large spherical volume, it seems plain that on the whole they do not participate in whatever rotation the galactic disk may have. The distribution of the clusters thus should be more or less stationary with respect to the center of the galaxy. Specifically, it is thought the rotational velocity of the system of globular clusters about the center of the Milky Way can be no more than 50 kilometers per second. Hence the rotational velocity of the sun about the center of the galaxy should be no more than 230 kilometers per second, and certainly no more than 250.

Meanwhile other investigators had determined the velocity of the sun with respect to the average motion of nearby galaxies, namely those whose distance from the sun is no greater than a million parsecs. Their result for the rotational velocity of the sun about the center of the Milky Way was 300 kilometers per second. The difference between the two results could best be interpreted as indicating that the center of the Milky Way galaxy is moving at a velocity of 50 to 80 kilometers per second with respect to the nearby galaxies. That velocity seemed surprisingly large.

Einasto argued that the result might nonetheless be correct and that the unsuspected massiveness of the Milky Way system might be the cause of it. To test his hypothesis he examined the motion of the Milky Way within the group of galaxies of which it is a member. Einasto proposed that there are subgroups within the group and that in one of the subgroups the Milky Way is gravitationally dominant over several other aggregations of stars, such as the two small nearby galaxies called the Large and Small Clouds of Magellan and a number of dwarf spheroidal galaxies, of which seven are now known. One of the dwarf spheroidal galaxies lies some 150,000 parsecs from the center of the Milky Way. The velocity of the sun with respect to the average motion of these galactic companions proved to be almost 300 kilometers per second. It too was surprisingly high. Einasto interpreted the velocity as an effect of the great

mass of our galaxy on its nearest neighbor galaxies.

Einasto was therefore strengthened in his suspicion that the Milky Way is more extended and more massive than had been supposed. In 1976 he offered a model of the Milky Way system in which the mass of the central bulge, the disk and an extended halo is 900 billion solar masses. Even that mass is insufficient to account for the great velocities observed among the galaxy and its companions. Einasto therefore proposed that the bulge, the disk and the halo are embedded in a still larger but nonetheless unseen component of the galaxy, the corona (as he called it), which extends out to at least 100,000 parsecs from the center and has a mass of 1.2 trillion solar masses. The total mass of the Milky Way would then be 2.1 trillion solar masses, or at least seven times the value accepted in 1975.

Supporting evidence was forthcoming from several investigations. First, Vera C. Rubin, W. Kent Ford, Jr., and Norbert Thonnard of the Department of Terrestrial Magnetism of the Carnegie Institution of Washington examined the Doppler shifts of spectral lines in the light emitted by matter in the outer part of 17 galaxies. Each such shift is a displacement of a spectral line to a wavelength different from the one it would have if the source of the radiation were motionless with respect to the instrument that receives it. The investigators concluded that the outlying matter in each galaxy circles the center of the galaxy just as fast as the matter nearer the center. To put it another way, a rotation curve—a graph of rotational velocity as a function of distance from the center of the galaxy—was essentially a horizontal line for the outer part of each galaxy they studied.

The discovery of such flat rotation curves is remarkable. After all, the distribution of brightness in the optical image of a typical spiral galaxy leads one to infer that the galaxy's visible matter is concentrated toward the center and gets sparse at the periphery. From this concentration one can deduce that the outermost visible matter ought to be moving in response to forces analogous to

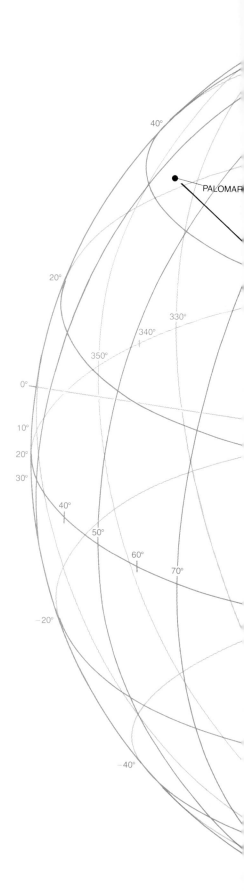

MAP OF THE MILKY WAY shows the galaxy in accord with the hypothesis that it is unexpectedly large and massive. The three long-recognized components of the galaxy lie well inside the coordinate grid. Their dimensions are given in parsecs. (A parsec is 3.26 light-years.) Among the three components the central bulge, with a radius of 4,000 to 5,000 parsecs, consists mostly of a dense packing of old stars. The galactic disk, with a radius of 15,000 parsecs, consists of younger stars and dust and gas. Its spiral features (*colored curves*) have been traced only in the sun's vicinity. The galactic halo, with a radius of 20,000 parsecs, consists mostly of a thin packing of old stars and of roughly half of the stellar aggregations called globular clusters. The hypothetical outermost component of the galaxy is called the corona; its presence is inferred from the velocities of the visible matter. Presumably the objects in the corona are not highly luminous. In this map the "galactic companions" visible in the corona to a distance of 100,000 parsecs are plotted by sun-centered coordinates. The direction from the sun to the center of the galaxy defines zero degrees galactic longitude. Angles above and below the galactic plane are measured in galactic latitude. The companions include 10 globular clusters, four dwarf spheroidal galaxies and the irregular galaxies the Large and Small Clouds of Magellan.

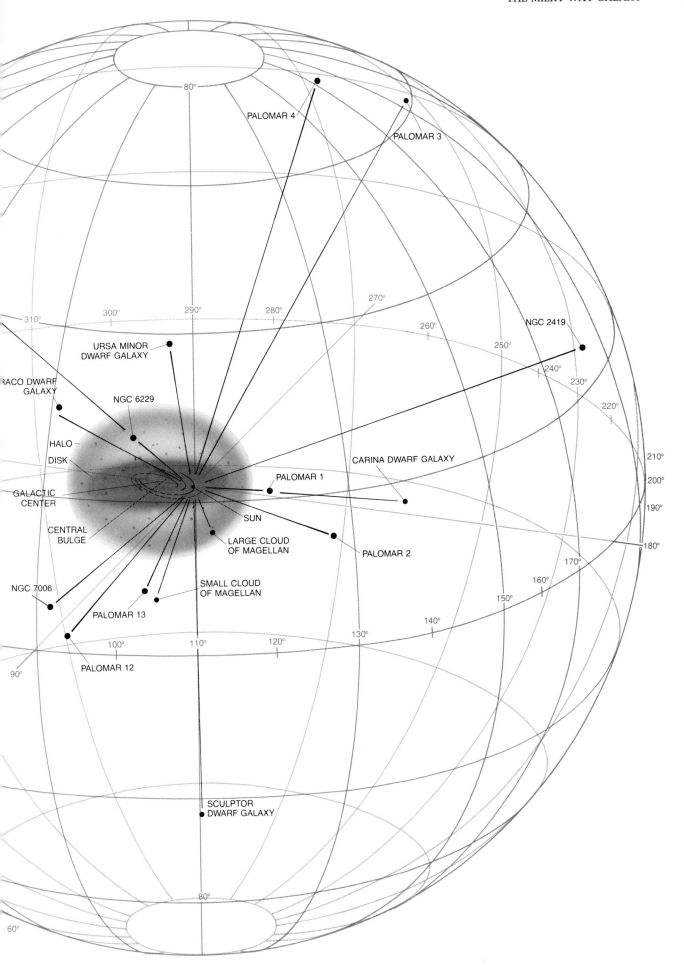

80°

PALOMAR 4

PALOMAR 3

270°

300°   290°   280°   260°   NGC 2419

310°   250°

URSA MINOR
DWARF GALAXY   240°

230°

RACO DWARF
GALAXY

NGC 6229   220°

210°

CARINA DWARF GALAXY

HALO   200°

DISK   PALOMAR 1

GALACTIC   190°
CENTER
                              SUN   180°

CENTRAL   LARGE CLOUD
BULGE   OF MAGELLAN   PALOMAR 2

                                                170°

NGC 7006   SMALL CLOUD   160°
                OF MAGELLAN
                                150°

PALOMAR 13   140°

                100°   110°   120°   130°

90°   PALOMAR 12

80°

60°

SCULPTOR
DWARF GALAXY

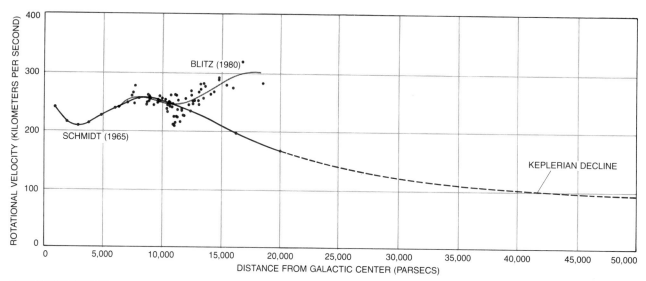

ROTATION CURVE graphs the circular velocity of matter in rotation about the center of the Milky Way. Here two such curves are drawn. A rotation curve plotted in 1965 by Maarten Schmidt of the Hale Observatories (*solid black line*) shows a circular velocity that declines toward the limit of the visible galaxy at 20,000 parsecs. If all the mass in the galaxy lay inside that limit, a test mass placed farther out would rotate at lower speed, in approximate obedience to a law first formulated (for the motion of the planets) by Johannes Kepler (*broken line*). Data analyzed by Leo Blitz and his colleagues at the University of California at Berkeley now yield a rotation curve (*colored line*) that rises toward a value of 300 kilometers per second at 20,000 parsecs. The rise of the newer curve implies unseen mass in great quantity outside the visible limit of the galaxy. Each point in the newer data is the circular velocity of a cloud of hydrogen atoms.

those acting on the outermost planets of the solar system. The peripheral matter ought to be rotating about the center of the galaxy at a lower circular velocity (expressed in kilometers per second) than the matter nearer the center. Evidently it does not. The rotation curves calculated by Rubin, Ford and Thonnard thus imply the presence of unseen matter in large quantity beyond the apparent periphery of each of the galaxies.

The Corona of the Milky Way

Studies similar to those of Rubin, Ford and Thonnard bear directly on the distribution of mass in the Milky Way. In one of them F. D. A. Hartwick of the University of Victoria and Wallace L. W. Sargent of the California Institute of Technology calculated the velocities of globular clusters at distances greater than 20,000 parsecs from the center of the galaxy. In other studies James E. Gunn, Gillian R. Knapp and Scott D. Tremaine employed data gathered at the Owens Valley Radio Observatory to determine the velocities of clouds of interstellar hydrogen atoms. Maurice P. Fitzgerald of the University of Waterloo, together with Peter D. Jackson of the University of Maryland and Anthony Moffat of the University of Montreal, determined the velocities of stars and star clusters as much as 17,000 parsecs from the center of the galaxy. William L. H. Shuter of the University of British Columbia determined the velocities of clouds of hydrogen and carbon monoxide that lie in the galactic anticenter: the direction opposite to the line of sight

from the earth to the center of the galaxy. Recently Leo Blitz and his colleagues at the University of California at Berkeley have measured the Doppler shifts in both the optical and the radio parts of the electromagnetic spectrum for lines from 184 nebulas and large clouds of interstellar hydrogen and carbon monoxide in the anticenter. All the results are consistent with a rotation curve that does not fall.

From the evidence available today it seems fair to conclude that the rotation curve for the Milky Way attains a value of 230 kilometers per second at 8,500 parsecs, the distance from the galactic center that marks the position of the sun. From there the rotational velocity continues to increase. It reaches 300 kilometers per second at a distance of 20,000 parsecs.

Several objects more distant than 20,000 parsecs are visible; a list of them was published by the International Astronomical Union in 1979. Four globular clusters lie between 20,000 and 40,000 parsecs from the galactic center; the Large Cloud of Magellan and two globular clusters lie between 40,000 and 60,000 parsecs; two dwarf spheroidal galaxies and the Small Cloud of Magellan lie between 60,000 and 80,000 parsecs, and one dwarf spheroidal galaxy and three globular clusters lie between 80,000 and 100,000 parsecs. Four more dwarf spheroidal galaxies and two globular clusters lie between 100,000 and 220,000 parsecs, but their claims to membership in the corona of the Milky Way are more dubious.

Plainly our galaxy is far more extend-

ed and of much greater mass than was hitherto thought; the Milky Way has been elevated to the rank of a major spiral galaxy. Taken together, however, the visible constituents of the corona have only a tiny fraction of the hundreds of billions of solar masses that ought to be there. Apparently the Milky Way shares with the galaxies studied by Rubin, Ford and Thonnard the property that much of its outlying matter is dark. Indeed, John N. Bahcall and Raymond M. Soneira of the Institute for Advanced Study infer from the invisibility of the hypothetical matter that if there are stars in the corona that are not bound into clusters or into dwarf galaxies, their intrinsic brightness should be less than a thousandth that of the sun.

What, then, is the unseen mass? Three facts are worth noting. First, dwarf galaxies and globular clusters consist mainly of old stars. Second, old stars are not highly luminous. Third, no one has detected from the corona the spectral lines that characterize clouds of gaseous matter such as hydrogen and carbon monoxide in more central parts of the galaxy. At present, therefore, the best suggestion is that the corona of the Milky Way is composed mainly of old, burned-out stars. On the other hand, the unseen mass of the galaxy's corona may not fit any of the categories based on what can be seen in more accessible regions. We do not know yet what is out there.

The Central Bulge

It may come as a surprise that the center of the Milky Way is no less mysteri-

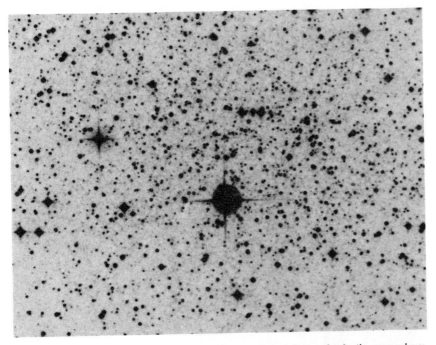

CARINA DWARF GALAXY is the latest addition to the catalogue of galactic companions; its estimated distance of some 100,000 parsecs from the center of the Milky Way apparently makes it an outer member of the galactic corona. In this negative print of a photograph the dwarf galaxy is the loose, inconspicuous aggregation of small dark points near the center. The larger points are stars much closer to the solar system along the same line of sight. The photograph was made as part of the United Kingdom–European Southern Observatory Sky Survey.

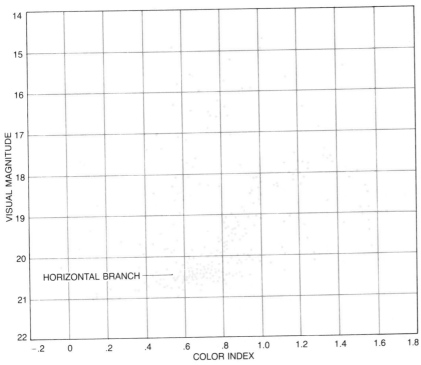

HERTZSPRUNG-RUSSELL DIAGRAM for stars in the central part of the Carina dwarf galaxy makes it possible to determine the distance of that galactic companion. The diagram plots the apparent luminosity of each such star (*vertical axis*) against a measure of its color (*horizontal axis*). The stars in what is called the horizontal branch of the diagram turn out to have an apparent luminosity of approximately 20.4. Their intrinsic luminosity would place them 20 magnitudes higher. From the dimming of their luminosity the distance of 100,000 parsecs is inferred. At a galactic latitude of −22 degrees, the Carina dwarf galaxy is well outside the central plane of the Milky Way; hence the correction of the calculation for the absorption of light by intervening interstellar matter should be small. The data were gathered by Russell Cannon and his associates at the Royal Observatory in Edinburgh and at the Anglo-Australian Observatory. The limit of sensitivity of the workers' apparatus is just below the horizontal branch.

ous than the galactic corona. In fact it is only slightly more visible to observers on the earth. Twenty-five years ago, when observations could be made only at visible and radio wavelengths, three kinds of object were known to be plentiful in the galactic center. Each of them is old. The first is globular clusters. The second is RR Lyrae variables. These are old stars that alternately brighten and dim with a period on the order of a day. The third is planetary nebulas. Most of them are the old, collapsed stars called white dwarfs, each one surrounded by a cloud of gas that is thought to be the shed atmosphere of the star. The galactic center itself was hidden from optical view by layers and shells of dust that are estimated to dim the light from the center by as much as 30 astronomical magnitudes, or a factor of 10^{12}. The dust is particularly effective at blotting out the blue light to which the photographic emulsions employed in earlier days of astronomical photography were often made most sensitive. Some of the dust lies in a lane at the margin of the central bulge, a few thousand parsecs from the solar system. Some of it lies only 300 parsecs from the solar system, in the constellation Ophiuchus, along the line of sight from the earth to the center. The only thing then known about the center was that it harbored a strong radio source. Electromagnetic radiation in the radio part of the spectrum is not absorbed by the dust.

Some 15 years ago the prospects began to brighten. The detection of radio emissions at a wavelength of 21 centimeters served to delineate the sources of such radiation: clouds of neutral (unionized) hydrogen atoms. The detection of emissions at other wavelengths revealed the presence of dark interstellar clouds consisting mostly of molecules. Above all, the advent of infrared astronomy opened a window to the center of the galaxy at wavelengths of from a micrometer to a millimeter. Infrared radiation constitutes a second band of wavelengths that the interstellar dust does not heavily absorb. One of the most useful new techniques is the detection of the infrared radiation emitted at a wavelength of 12.8 micrometers by ionized neon atoms. Because neon is a by-product of energetic events such as stellar explosions it becomes a ubiquitous constituent of interstellar clouds. The neon therefore serves as a tracer.

The observations at radio and infrared wavelengths yield four kinds of data. First, they reveal the presence of local maximums in the rain of radiation from certain directions in the sky. Second, the Doppler shift of spectral lines in the radiation reveals the radial velocity of such a source: the speed of the source either toward the solar system or away from it. Third, the broadening of a spectral line can suggest that the source

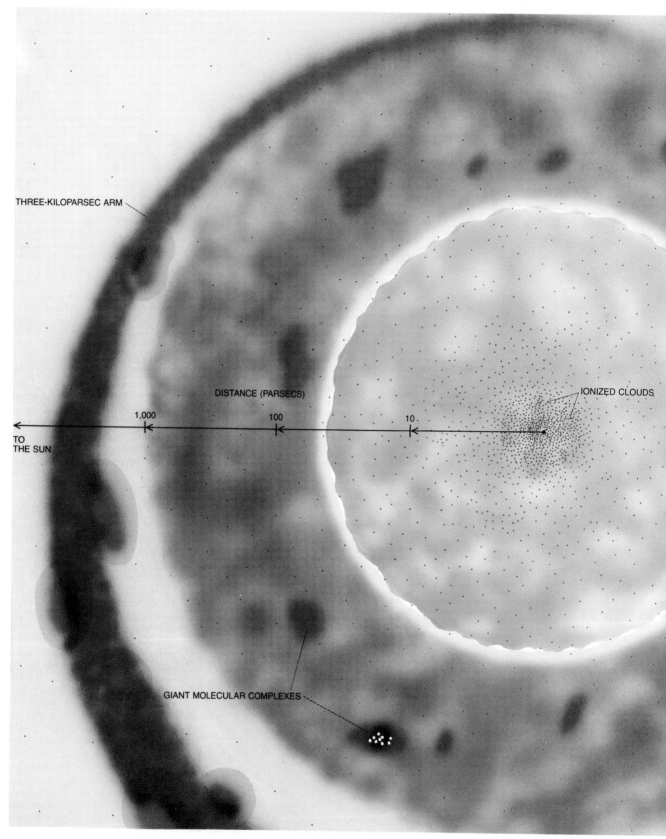

CENTRAL PART of the Milky Way is shown schematically in an illustration devised by Thomas R. Geballe of the Hale Observatories. The view is from above. The scale decreases logarithmically with distance from the center, so that the innermost parsecs are magnified. The three most central parsecs include the densest congestion of stars (*colored dots*) in the galaxy. (Throughout the illustration the density of stars has been reduced by a factor of 2,000.) The region also harbors compact clouds of ionized gas (*dark color*) that are made up mostly of hydrogen. The velocity of the clouds suggests they are circling a supermassive object, perhaps a black hole, that lies precisely at the center. A thinner distribution of ionized gas (*light color*) permeates the central 100 parsecs. It is surrounded in turn by a ring of cooler, unionized gas (*gray*) whose hydrogen consists of both atoms and molecules. The ring includes giant complexes of dust and molecules (*dark*

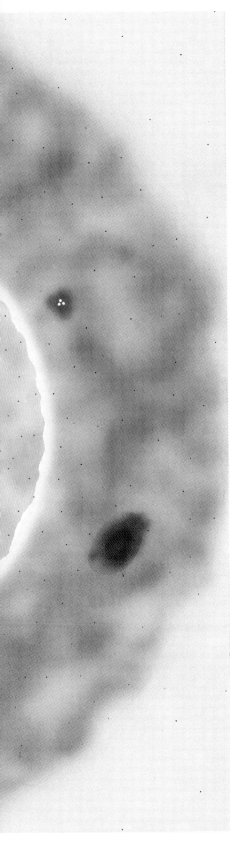

gray). **In some of them young stars (*white dots*) have formed. A dense band made up mostly of un-ionized hydrogen appears at the left. It is the innermost part of an expanding feature called the three-kiloparsec arm. On one hypothesis the arm was created by an explosion at the galactic center 30 million years ago.**

is expanding or contracting. Fourth, the relative intensities of certain lines in the spectrum suggest the temperature of the source. Even when all possible information has been extracted from the radiation, however, much remains uncertain. It cannot be ascertained, for example, whether a source in approximately the direction of the center of the galaxy lies in front of the center or behind it, or whether the radial velocity of a source actually means it is rotating about the center.

Let me summarize the present state of knowledge about the central bulge of the Milky Way. In its overall shape the central bulge is a slightly flattened sphere. Its outer boundary, which lies 5,000 parsecs from the center, is marked by a ring of what are now called giant molecular complexes. They are large, dark, clumpy interstellar clouds made up mostly of hydrogen molecules. I shall be returning to them. The bulge itself consists in general of a dense clustering of old stars in a rather thin matrix of interstellar gas and dust. The stars are known from their infrared radiation, which can be distinguished from that of gas or dust. One way to account for the relative scarcity of gas and dust in the bulge is to assume that much of it placidly condensed long ago to form the stars of the bulge. On the other hand, several interstellar features in the bulge suggest that the center of the galaxy has had a complex and violent history.

The outermost detectable feature of the bulge is a ring of neutral hydrogen at a distance of 3,000 parsecs from the center. The ring was discovered in 1964 by Jan H. Oort and G. W. Rougoor of the Leiden Observatory. The Doppler shifts of the radiation it emits show it is rotating and, more important, expanding, with velocities away from the center that range from 50 to 135 kilometers per second. Perhaps the ring is a new spiral arm unfurling. One is equally tempted, however, to speculate that the center of the galaxy expelled a kind of smoke ring some 30 million years ago. It is as if there had been a titanic explosion there. Perhaps the explosion swept away much of the gas and dust in the bulge.

The next feature inward lies at a distance of about 1,500 parsecs from the center. It is construed by its discoverers, Butler Burton of the University of Minnesota and Harvey S. Liszt of the National Radio Astronomy Observatory, to be a disk of both atomic and molecular hydrogen. It too is both rotating and expanding. Surprisingly, the best way to interpret the data is to assume that the disk is tilted at an angle of 15 to 20 degrees to the plane of the galaxy.

One might hope that the composition of the central bulge would be more or less homogeneous from the Burton-Liszt ring to the center, if only to simpli-

fy the task facing those who attempt to understand the structure of the bulge. More surprises, however, await. Another smoke ring is evident some 300 parsecs from the center. This one too is a mixture of molecular complexes, dust clouds and regions of atomic and molecular hydrogen. The atomic hydrogen is ionized in places, which means it is quite hot: well over 10,000 degrees Kelvin. Associated with these hot spots are clusters of newly formed blue-white supergiant stars. Why should these realms of high temperature and star formation lie precisely in this ring? One is particularly puzzled because a cooler ring of only mildly ionized atoms at a temperature of 5,000 degrees lies a mere 10 parsecs from the center. The 10-parsec ring is rather dense, and it is rotating.

The central three parsecs of the galaxy evidently includes several million stars; they give the center the densest packing of stars in the galaxy. The core region also includes a number of compact clouds of ionized gas; a group of workers led by John H. Lacy and Charles H. Townes of the University of California at Berkeley has detected 14 of them. A typical cloud has about the mass of the sun and a diameter of a fraction of a parsec, and it is speeding around the center: it completes an orbit in some 10,000 years (compared with the sun's orbital period of some 200 million years). Luis Rodriguez and Eric J. Chaisson of the Harvard College Observatory have shown that the velocity of the ionized gas increases with proximity to the center. All of this suggests the clouds are satellites of a supermassive innermost object.

Whatever it is at the very center appears as a bright infrared source in maps made by Eric E. Becklin and Gerry Neugebauer of the California Institute of Technology. According to Bruce Balick of the University of Washington and Robert L. Brown of the National Radio Astronomy Observatory, who have studied the radio emissions of the central object, it has a diameter no greater than 10 times the distance from the earth to the sun. Its mass may be as great as 50 million solar masses. The most likely conjecture is that the very center of the Milky Way harbors a black hole created by the infalling of hundreds of thousands of stars. The center would then be in essence a stellar graveyard.

Optical Spiral Structure

When the periphery and the center of the Milky Way have been considered, there remains a middle region, which has the solar system in its midst. It is the part of the galaxy in which spiral structure prevails.

The tracing of the spiral arms of the Milky Way began in earnest three decades ago, when Morgan, Osterbrock and Sharpless distinguished three spiral-arm segments. To trace them they plotted the positions of the blue-white supergiant stars classified on the basis of their pattern of spectral lines as *O* and *B* stars, together with the bright clouds of ionized hydrogen atoms that often surround such stars. Their diagrams show an Orion arm, of which the sun is a member; a Perseus arm, 2,000 parsecs farther from the center of the galaxy,

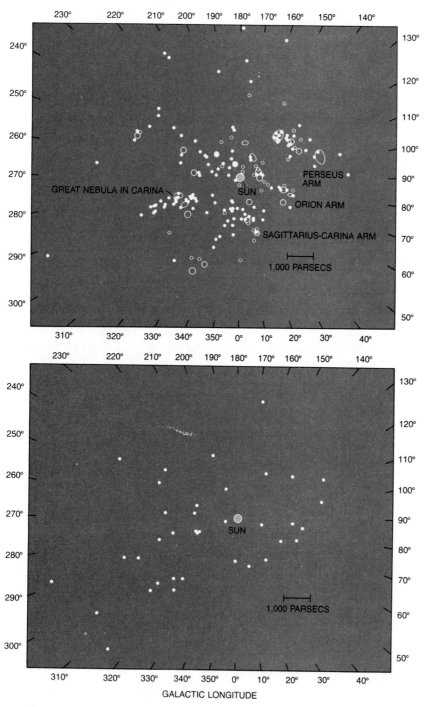

and a Sagittarius arm, 2,000 parsecs closer to the center of the galaxy. The naming of the arms reflects the general practice in astronomy of employing the constellations to signify directions in the sky. To the three arms distinguished by Morgan and his colleagues has since been added the Carina arm, which may be a continuation of the Sagittarius arm; the concatenation is called the Sagittarius-Carina arm. The two segments that compose it meet at the Great Nebula in Carina, which enmeshes a large number of *O* and *B* stars. There is evidence for other segments as well.

Recent work by Roberta M. Humphreys of the University of Minnesota confirms that *O* and *B* stars are abundant in the principal arms that were recognized three decades ago. This is cheerful news. It means the spiral arms are indeed delineated by very hot, blue-white supergiant stars, by the clusters made up of such stars and by the bright clouds of gas in which such stars and clusters are found. *O* and *B* stars are quite young; the ones seen now in the spiral-arm segments of the Milky Way were formed no more than 10 million years ago. There is no question the spiral arms are regions of star formation.

The inner margins of most spiral arms are marked, it seems, by dark nebulas: cold clouds of atoms, molecules and dust. According to the density-wave theory of Lin and Shu, the dark nebulas signal the compression of the interstellar matter by a wave of pressure that advances through the galactic disk. The compression precipitates the condensation of the matter into the stars of the spiral segment. The young stars might be expected to lie behind the advancing density wave and dark nebulas to lie inside it. According to the theory, however, the wave moves at only two-thirds the speed of the galactic disk's rotation. Hence the wave is overtaken by the stars that formed inside it, and the loci of dark nebulas come to lie at the trailing edge of the stars.

Certain of Humphreys' results, it should be said, are less than cheerful. Some of the stars called Cepheid variables are as young as *O* and *B* stars; in particular the Cepheid variables that alternately dim and brighten with a period longer than 15 days have ages of no more than 10 million years. Moreover, they too are large and bright, and so they can be seen at great distances. Since they are young, they should not have moved far from their birthplace: presumably a dark nebula at the inner edge of a spiral arm. All things considered, the long-period Cepheid variables should be excellent delineators of spiral structure. They are not. The Cepheid variables plotted by Humphreys form an essentially random distribution of points in the galactic plane.

One thing must always be kept in

SPIRAL STRUCTURE of the disk of the Milky Way in the vicinity of the sun has been traced by observations at optical wavelengths with varying success, depending on what classes of astronomical objects are employed in the attempt. The upper chart shows the positions of loose groups of young stars (*open circles*) and of clusters in which at least some of the stars are young (*solid circles*); the distribution of such stars confirms the presence of the spiral features called the Perseus arm, the Orion arm and the Sagittarius-Carina arm. The lower chart shows the positions of long-period Cepheid variables. These young giant stars are thought to form inside spiral segments. Although the same part of the galaxy is mapped, no spiral structure appears. Data for the charts were collected by Roberta M. Humphreys of the University of Minnesota.

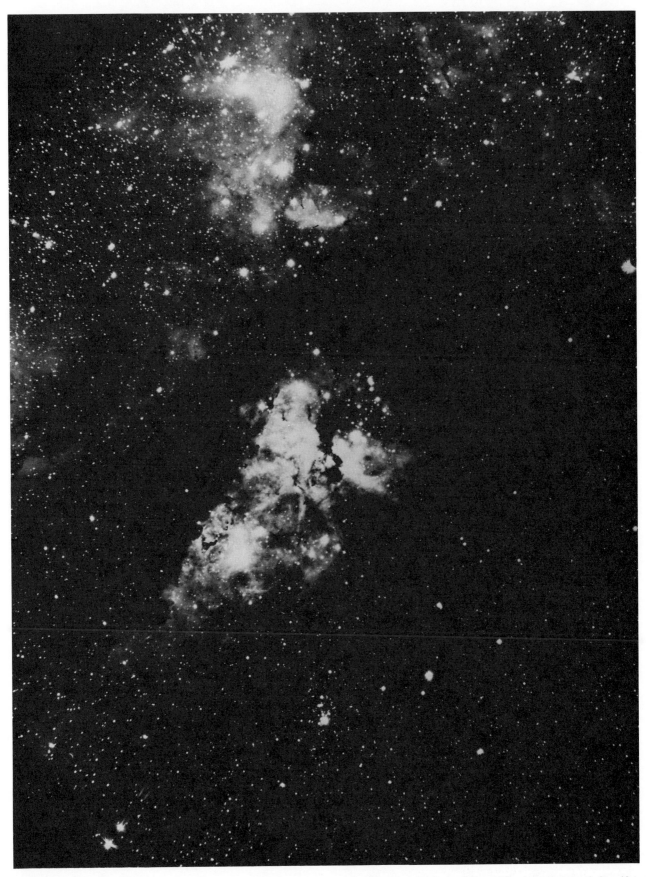

BRIGHTEST SPIRAL FEATURE in the vicinity of the sun is the Great Nebula in Carina, which marks the place where the Sagittarius arm and the Carina arm meet. The nebula is a cloud of hydrogen that has been ionized, and hence rendered luminous, by the ultraviolet radiation of newly formed blue-white supergiant stars in its midst. The nebula cannot be seen from the Northern Hemisphere. The distance to the nebula is 2,700 parsecs. The photograph was made at Mount Stromlo Observatory of the Australian National University.

mind by those who search for spiral structure in optical observations of the Milky Way. The usual method of calculating the distance to a star is to compare its observed luminosity with what its intrinsic luminosity is taken to be. The intrinsic luminosity is assigned on the basis of the pattern of lines in the spectrum of the star's radiation. The diminution of the brightness is a correlate of the distance. Even with the best techniques, however, the distance of an O star or a B star is uncertain by plus or minus 10 percent. For example, a star that is calculated to be 2,700 parsecs from the

sun—say a star in the Carina nebula—could be as little as 2,400 or as much as 3,000 parsecs away. The result is a purely observational blurring of several hundred parsecs in the plotting of what might actually be a spiral feature.

With the most modern telescopes, spectrographs and photographic equipment, O and B stars can be detected, and their spectra can be examined, at distances calculated to be as great as 8,000 parsecs from the sun. At that distance, however, the uncertainty in the calculation is plus or minus 800 parsecs. In such a case a spiral feature may well be un-

recognizable. There seems little prospect of tracing the spiral structure of the Milky Way by optical means to distances beyond 8,000 parsecs.

Radio Spiral Structure

How do the efforts to trace spiral structure fare at radio wavelengths? Again the prospects are far from encouraging. By the early 1950's, when Morgan, Osterbrock and Sharpless presented the key optical features of the spiral structure, Harold I. Ewen and Ed-

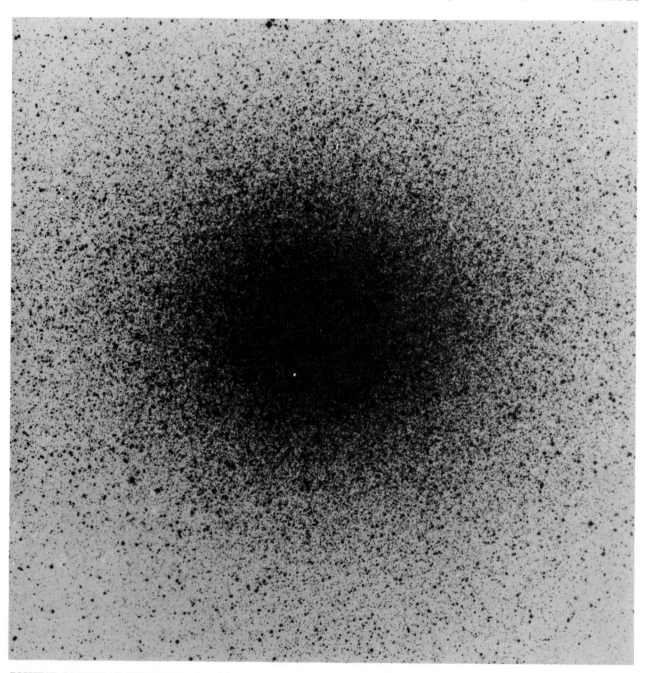

RICHEST GLOBULAR CLUSTER in the vicinity of the sun is Omega Centauri, shown here in a negative print of a photograph made by Gary S. Da Costa with the 2.5-meter telescope of the Las Campanas Observatory in Chile. The cluster includes several hundred thousand stars within a diameter of approximately 30 parsecs. Omega Centauri is slightly more than 5,000 parsecs from the solar system.

ward M. Purcell of Harvard University had detected radiation from interstellar clouds of neutral atomic hydrogen at a wavelength of 21 centimeters. Within a few years the first radio maps of the galaxy were available. They suggested that at least in the outer parts of the galactic disk the clouds of neutral atomic hydrogen are arrayed in nearly circular spiral features.

Some details of the way in which such data are amassed and interpreted are worth examining. When a radio telescope detects radiation at and near a wavelength of 21 centimeters from a small area of the sky, it is actually receiving radiation from a number of clouds of neutral atomic hydrogen along a single line of sight. Each cloud has its own velocity of approach or recession with respect to the telescope, and hence its radiation has a distinctive Doppler shift. As a result the telescope receives from a single direction a profile of peaks and valleys in a graph of intensity v. wavelength. The telltale 21-centimeter line is actually a set of closely spaced lines with various intensities, various degrees of broadening and various Doppler shifts. Since neutral atomic hydrogen is assumed to have its greatest density at the trailing edge of spiral features, it seems reasonable to assume that a peak of great strength in the profile should lie at the wavelength corresponding to the velocity of approach or recession of hydrogen where the line of sight crosses a spiral feature. With the aid of a rotation curve for the Milky Way one should then be able to determine the distance to the feature.

This chain of reasoning is simple and lovely, but it does not stand up under scrutiny. The first complication is that the clouds of neutral atomic hydrogen have motions of their own, quite apart from the motion of the spiral features. These independent motions can easily alter the total velocity of a cloud by as much as six kilometers per second. An additional complication is that the matter of the galactic disk apparently exhibits large-scale streamings. Burton and his associates at the National Radio Astronomy Observatory have demonstrated that even slight motions of this kind can give rise to peaks of intensity in the 21-centimeter radiation for directions in which the line of sight does not cross a spiral feature. Further still, there are directions in the sky for which the velocity of approach or recession of a cloud of gas may change only slowly with distance. The 21-centimeter profile may then show a peak caused by contributions from the width of a single great expanse of gas of uniform density.

Several other approaches can be made to the study of spiral structure. One is to observe the spectral lines emitted by molecules of carbon monoxide at

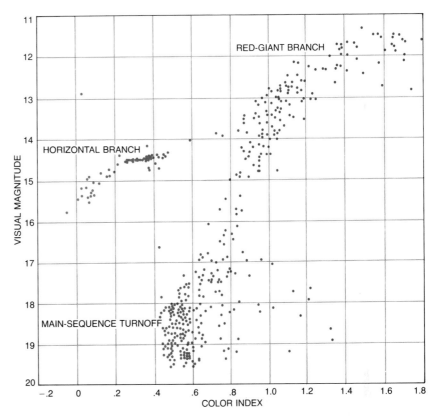

CURIOUS STELLAR COMPOSITION of the Omega Centauri globular cluster is suggested by a Hertzsprung-Russell diagram for the stars that make up the cluster. The branch of the diagram consisting of red-giant stars (*upper right*) is unusually long. That signifies an unusual degree of variation in the giant stars' content of "metals," or atoms heavier than carbon. Specifically, the metal atoms in the star increase its opacity to the radiation escaping from inside. They thereby change its color: the metal-rich red giants are redder, and on a Hertzsprung-Russell diagram their position is well toward the right. The variation in metal content suggests in turn that the stars in the Omega Centauri cluster formed in more than a single episode. Further deductions are problematical. In general the age of a globular cluster is determined by comparing the luminosity and the color of the stars at what is called the main-sequence turnoff of the Hertzsprung-Russell diagram (*bottom*) with that of the stars at the corresponding position in diagrams generated in computer simulations of a cluster's evolution. In the case of Omega Centauri the data do not reveal the full extent of the main sequence, and so it is difficult to place the turnoff precisely. That makes it difficult to establish the age (or ages) of the cluster. The data were collected by Cannon and N. J. Stewart of the Royal Observatory in Edinburgh.

radio wavelengths close to 2.6 millimeters. The carbon monoxide lines give access to a class of clouds that are cooler than the ones composed of neutral atomic hydrogen. The cooler clouds are composed mostly of molecular hydrogen, with some admixture of carbon monoxide and other substances. (The general principle is that increasing temperature leads first to the dissociation of molecules into neutral atoms and then to the ionization of the atoms.) William Herbst of Wesleyan University has shown that the presence of carbon monoxide in an interstellar cloud is correlated with the presence of dust. The carbon monoxide lines therefore aid in the detection of distant dust clouds and dust complexes that might otherwise escape notice.

A low-resolution survey of carbon monoxide clouds by Richard S. Cohen and Patrick Thaddeus of the Goddard Institute for Space Studies and Thom-

as M. Dame of Columbia University shows evidence of the existence of spiral features. Indeed, in some instances the carbon monoxide clouds seem to fill gaps between clouds of atomic hydrogen in some of the recognized spiral features. Still, I doubt the spiral structure of the Milky Way would have been discovered if the only data available had been the radio observations at a wavelength of 2.6 millimeters.

I thoroughly dislike having to be pessimistic about the prospects for mapping the spiral structure of the Milky Way beyond a distance of about 8,000 parsecs from the sun, particularly since the structure seems firmly established up to that distance, but I see no chance for improvement in the next decade or two. This is not to say there are now no possibilities whatsoever for the study of spiral structure. Among optical telescopes the great reflectors already in place and the Space Telescope, an in-

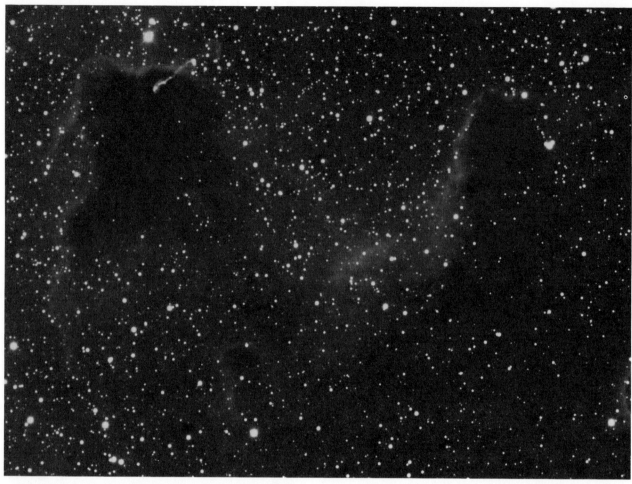

BIZARRE BIRTH OF STARS is visible in the small dark cloud at the left in this photograph. The cloud is one of those called a globule. It is half a parsec in diameter and probably has somewhat less than 100 times the mass of the sun. The mass of the globule is composed of molecules (mostly hydrogen) and dust. From this particular globule a pair of incipient stars connected by a luminous strand have evidently been expelled. The strand crosses the upper edge of the cloud. The expulsion defies the hypothesis that a globule condenses in about a **million years to form a single new star. The contour of brightness extending diagonally from the bottom toward the right of the field is an edge of the Gum nebula, which is thought to be in part the debris of a supernova explosion. Both the globule and the nebula are calculated to be approximately 300 parsecs from the solar system. They lie in the part of the southern sky marked by the constellation Vela. The photograph was made by the author in 1978 with the four-meter telescope at the Cerro Tololo Inter-American Observatory in Chile.**

strument that will orbit the earth, promise to reveal the fine details of spiral structure and the motions of small-scale features. Among radio telescopes the Westerbork Array in the Netherlands and the Very Large Array in New Mexico, which synthesize images from a set of detectors, will advance the investigation of the cold, dark clouds in spiral features and no doubt will suggest much about the causes of spiral structure. These promising developments, however, apply mainly to spiral galaxies other than our own. The spiral galaxies Messier 31, 33, 51, 81 and 101 are a few fairly nearby examples.

The Ages of Globular Clusters

One realm of research that is flourishing in both galactic and extragalactic astronomy is the examination of globular clusters. The clusters are important in part because they seem to be the oldest

objects in the Milky Way. They therefore hint at the birth and evolution of the galaxy and indeed of the early universe. The simplest hypothesis about the clusters is that they all formed within a short time (say a billion years) of the big bang: the instant when all the matter in the present universe emerged explosively, it is thought, from a single point. The clusters would then have been among the first objects to condense as the galaxies took shape, each cluster evolving from a large blob of gas. It counts in favor of this hypothesis that half of the roughly 200 globular·clusters in the Milky Way lie scattered throughout the almost spherical volume of the galactic halo. Presumably they formed there well before the galactic disk took shape. The orbit of such a cluster is typically a rather eccentric ellipse and not the more nearly circular orbit characteristic of matter in the disk. The major axis of the orbit is sometimes several tens of thou-

sands of parsecs, and once every billion years the orbit sends the cluster rushing through the thickness of the disk, which is only a few hundred parsecs.

On the simplest hypothesis the stars of the globular clusters would have formed at a time when the available matter in the galaxy was mainly hydrogen and helium, the two chemical elements presumed to have been created in the immediate aftermath of the big bang. In contrast, a later-born star would condense from interstellar gas part of which had been cycled through the interior of the first-born stars. Some of the gas, for example, would have been propelled into space by supernova explosions. It would be matter in which heavier atoms had been created by thermonuclear fusion. The later-born star would therefore have higher concentrations of the heavy chemical elements; in the shorthand of astrophysicists all such elements are known as metals.

It is the concentration of metals in the various globular clusters that imperils the simplest hypothesis. To be sure, the ratio of metals to hydrogen and helium is 100 times greater in the sun than it is in the stars of metal-poor globular clusters such as M3 (for Messier 3). Moreover, the metal-poor clusters tend to be the ones that are outermost in the galaxy. An age of 15 billion years is now assigned to them. On the other hand, the globular cluster 47 Tucanae is relatively metal-rich. Its age is thought to be 10 billion years, which is twice the age of the sun. Omega Centauri, the most impressive globular cluster in the Milky Way, shows a range of metal concentrations. Evidently it is an idiosyncratic agglomeration of stars that were born at different times. The spread of ages for the globular clusters conflicts with current models of how the galaxy evolved. No model allows as much as several billion years for the galactic disk to have condensed.

With respect to the question of age it is notable that the maximum age of the universe can be inferred from the velocities at which galaxies are receding from one another. The current value for the maximum is close to 15 billion years. Recent investigations of the recession rate have tended to reduce the maximum age. If the trend continues, the age could conceivably fall to as little as 10 billion years. Then one would have to explain the finding that certain globular clusters seem to exceed the age assigned to the universe. It is clearly important that trustworthy values be calculated for the ages of the globular clusters. It would be particularly useful if such ages were available by the end of this decade, because by then the Space Telescope should have yielded far more reliable values for the recession velocities and the distances of the galaxies. It will thus have yielded a far more reliable value for the maximum age of the universe.

The Evolution of Globular Clusters

The internal dynamics of the globular clusters are well described by three characteristic times. The first is the crossing time: the time it takes a star to move across the cluster under the gravitational attraction of the cluster as a whole. The second is the relaxation time: the time in which the star settles down and becomes a stable member of the cluster under the influence of its gravitational interactions with nearby stars. The third is the evolution time: the time in which a stable cluster changes its form and its stellar composition significantly. For a globular cluster rich in stars the crossing time is much shorter than the relaxation time, which in turn is much shorter than the evolution time.

The work of Ivan R. King of the University of California at Berkeley has

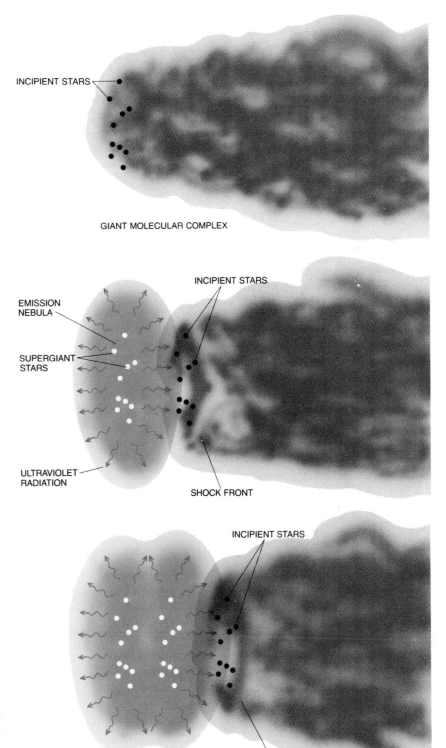

BIRTH OF SUPERGIANT STARS is thought to take place in repetitive stages. The sequence begins (*top drawing*) with the formation of a cluster of supergiant stars near the periphery of a giant molecular complex, a cold, dark region of molecules (mostly hydrogen) and dust. After a few million years the ultraviolet radiation of the new stars has ionized the nearby hydrogen, so that the stars are enmeshed in an emission nebula: a cloud of luminous gas. In addition the pressure of the radiation has compressed the matter in the giant molecular complex. (Each quantum of the radiation has momentum, and so it pushes the molecules and the dust.) The result is the formation of a second group of supergiant stars (*middle drawing*). In another few million years a similar sequence may form still another group of stars (*bottom drawing*). The diagram is based on a hypothesis put forward by Bruce G. Elmegreen of Columbia University and Charles J. Lada of the University of Arizona. The hypothesis accounts for clusters of supergiant stars in nebulas, such as the Great Nebula in Orion, that lie next to a molecular complex.

suggested some further quantifications. When a star-rich globular cluster has been in existence for roughly 50 crossing times, its stars will have settled down; the statistics of their velocities will be much like those for the velocities of the molecules in a cloud of gas. This state of equilibrium might persist indefinitely if the stars never changed internally and if no stars left the cluster.

As early as the 1930's, however, Lyman Spitzer, Jr., of Princeton and Victor A. Ambartsumian of the Byurakan Astrophysical Observatory in Armenia had concluded that the stars of lowest mass in a cluster were most likely to have the greatest velocity, and that in many cases the stars with the greatest velocity would escape from the cluster. Because of their great velocity the escaping stars would take with them more than the average share per star of the total energy of the cluster. As a result the cluster would contract. Over a period of time on the order of the evolution time the cluster would lose appreciably in average energy per star, and the most massive stars would settle close to the center.

Internal changes among the stars that remain can only quicken this trend. To begin with, the loss of mass by stars is now recognized to be an astrophysical commonplace. Early in its evolution a typical star is embedded in a shell of gas that is gradually expelled by the out-

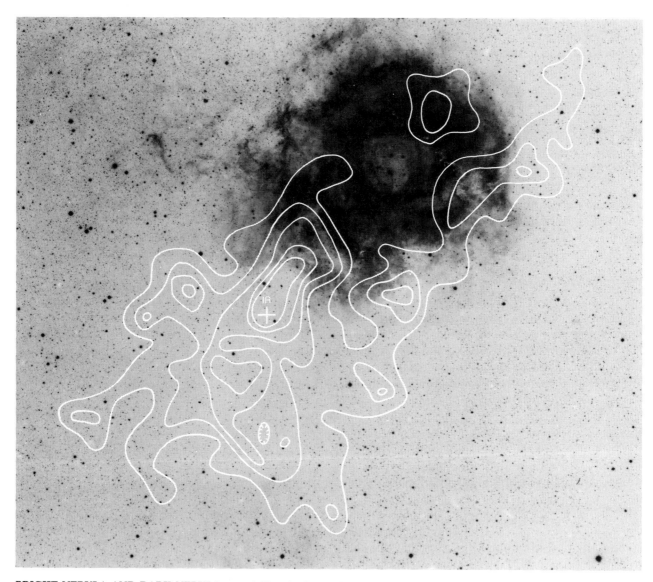

BRIGHT NEBULA AND DARK NEBULA are neighbors in the constellation Monoceros. Together they form an example of the mode of star formation diagrammed on page 84. The bright (or emission) nebula, called the Rosette, is the dark cloud at the upper right of this negative. Its hydrogen atoms have been ionized by the ultraviolet radiation of a group of young blue-white supergiant stars. Several such stars lie in what appears to be a hole at the center of the nebula. The radiation from the star precisely at the center is particularly intense and may have blown out the hole. The dark nebula is a giant molecular complex. It consists almost entirely of cold molecular hydrogen. Its full extent is best revealed, however, by the radiation of carbon monoxide molecules at a radio wavelength of 2.6 millimeters. The contour lines in the illustration map antenna temperature, which indicates the strength of the carbon monoxide radiation. The pattern of the contours suggests that the ultraviolet radiation of the supergiant stars is eating into the complex. A shock front along which the matter in the complex is compressed by the radiation apparently lies at the lower left edge of the Rosette. The front is presumably a place where new stars will condense. One peak of the carbon monoxide radiation corresponds to a source of infrared radiation (*IR*) now identified as a newborn star embedded in gas and dust. Both the Rosette nebula and the giant molecular complex are 1,600 parsecs from the solar system. The illustration displays the results of a study of the interaction between the nebulas by Leo Blitz of the University of California at Berkeley and Patrick Thaddeus of the Goddard Institute for Space Studies. The photograph is from the Palomar Sky Survey.

ward-directed pressure of the star's electromagnetic radiation. (Each photon, or quantum of the radiation, has momentum, and so it pushes whatever absorbs it.) Later the star begins to shed a gentle stellar wind of particles. Still later certain kinds of star explode. In a nova the exploding star ejects into space the equivalent of the mass of a planet such as Jupiter. A supernova is an explosion in which an entire star is destroyed. Because mass is lost through mechanisms such as these the stars in a globular cluster move systematically to classes with lower mass (and also fainter intrinsic brightness). Over a period of time on the order of the relaxation time they acquire the higher velocity of a less massive star. Hence their chance for escape from the cluster improves.

Throughout the hundreds of billions of years a cluster may spend in the thin outer halo (or even the corona) of the Milky Way the evolution of the cluster is affected only by the internal events I have just described. There comes a time, however, when the cluster traverses the galactic disk. At such a time the cluster is at risk not so much of collision as of gravitational interaction with the matter in and near the galactic plane. The force of the interaction may tear the cluster apart.

As the orbit of a cluster brings it close to the center of the galaxy a further threat arises. The threat exists because the central mass of the galaxy exerts a greater attractive force on the side of the cluster that passes nearest the center than it does on the side farthest away. The difference between the inner and the outer force can deform the cluster. Indeed, for any globular cluster there is a critical radius: a distance from the galactic center within which some of the stars in the cluster will be sheared off as the cluster makes its closest approach to the center. The critical distance is called the tidal radius because the mechanism that deforms the cluster is similar to the one that raises tides in the oceans. It is not surprising that the globular clusters inside the galaxy's central bulge are smaller on the average than the ones in either the galactic halo or the galactic corona.

One final factor may affect the globular clusters, but I mention it with some misgivings. As recently as 15 years ago few astronomers would have thought the formation of double star systems would be important in star clusters. Then Sebastian von Hoerner of the National Radio Astronomy Observatory made the first computer simulations of open clusters: aggregations of stars that differ from globular clusters in that the stars are young, fewer and more widely dispersed. Von Hoerner based the simulations on equations that represent the gravitational interactions among small numbers of stars. The simulations consistently showed that the stars in a cluster tend to unite by pairs into binary systems.

In more recent computer simulations incorporating first 100 and later 500 and even 1,000 interacting stars, the same effect was noted. In 1975 Douglas C. Heggie of the University of Edinburgh showed that a newly formed binary system is a formidable sink of the energy available in the cluster. In essence the formation of each binary system takes kinetic energy from the motion of the stars and converts it into energy that binds the two stars into orbit, each one around the other. The pair accordingly sink toward the center of the cluster. Still more recently, Spitzer and his associates at Princeton have devised models suggesting that in the crowded central region of a globular cluster, where the most massive stars are huddled, each star is likely to capture one or even two or more companions. As a result the central region may become even more densely packed as the globular cluster evolves.

The reason for my misgivings in the face of all this circumstantial evidence is simple. Massive binary star systems do indeed prevail in open clusters. Mizar is a prime example. It is a star in the Ursa Major open cluster. More recognizably, it is the star at the bend of the handle of the Big Dipper. Mizar has a companion called Alcor, which is faintly visible to someone with excellent eyesight. Evidently, then, Mizar is a binary system. Actually both Mizar and Alcor have companion stars. Each one is a binary system. That makes it all the more exasperating that the search for instances of massive binary star systems near the center of globular clusters has not been equally successful. In fact, it has had no successes at all.

Something else has been found, however. In 1976 Jonathan E. Grindlay and Herbert Gursky of the Center for Astrophysics of the Harvard College Observatory and the Smithsonian Astrophysical Observatory reported that an X-ray burst of terrific strength had reached the solar system from a direction close to that of the center of the globular cluster NGC 6624. The burst lasted for only eight to 10 seconds. Similar bursts have since been detected repeatedly from six globular clusters. According to one hypothesis, there is a black hole with a mass not less than 100 solar masses at the center of each such cluster. The bursts would then result from the intermittent collision of interstellar gas with a hot accretion disk of gas that surrounds the black hole. It comes to mind that the hypothetical black hole might be the ultimate consequence of the grad-

ual collapse of stars toward the center of the cluster, a collapse to which the formation of binary star systems may have made a contribution, and perhaps even given the final pushes.

Star Birth in Molecular Complexes

It was apparent even 35 years ago that the interstellar medium in the disk of the Milky Way includes clouds of gas and dust in which newborn stars are condensing. After all, some of the most striking objects in the galaxy are emission nebulas: bright clouds of ionized hydrogen atoms that are energized by the groups of giant stars inside them. The stars emit radiation at such a prodigious rate that they cannot have been doing so for more than a few tens of millions of years. The stars are therefore quite young. With only optical observations of the clouds, however, it was difficult to advance our understanding of how the stars actually formed.

All of this has now changed. Radio astronomy has revealed the presence in the clouds of more than 50 kinds of interstellar molecules, from hydrogen molecules, the lightest and by a factor of 1,000 the commonest, to a nine-carbon chain, the heaviest. Each species of molecule emits electromagnetic radiation at characteristic radio wavelengths. Moreover, infrared astronomy is now equipped to reveal the incipient stars themselves within their dense, obscuring clouds.

In almost all the processes in which a new star is thought to form, the initial step is the development of a concentration of matter—I shall call it a nodule—inside a cloud of interstellar atoms and molecules (for the most part hydrogen molecules), with a small admixture of dust. Some of the clouds are actually enormous clumpy distributions of matter. A giant molecular complex, for example, can have a mass of several hundred thousand solar masses. Other clouds are much smaller. Some of the clouds called globules have masses of only 20 solar masses. What all the nodules have in common is that they tend to collapse, mostly under the influence of their own gravitation, with perhaps occasional pressure from outside. An outside pressure that may be quite important is the one exerted on a cloud and the incipient nodules inside it when a density wave passes through and a spiral arm begins to form. The wave may well accelerate each of the processes I shall now describe.

The conditions in the roughly 4,000 giant molecular complexes that lie within 13,000 parsecs of the center of the galaxy seem ready-made for the formation of new stars. For one thing, each complex has plenty of matter. The mass

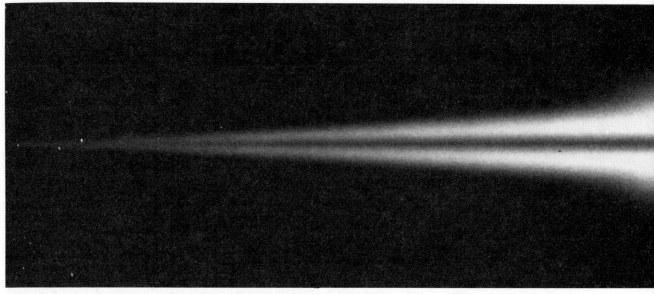

OUTSIDE VIEW of the Milky Way shows the galaxy edge on as it might be seen through a telescope by an astronomer in another galaxy. The view was generated with the aid of a computer by John N. Bahcall and Raymond M. Soneira of the Institute for Advanced Study.

of a typical complex is as great as several hundred thousand solar masses. Its diameter is about 50 parsecs. It is the most massive object in the galaxy (unless the galactic center has a supermassive black hole). The mass of a giant molecular complex is thought to be almost entirely hydrogen molecules; at a temperature of 20 degrees K., the cloud is too cold for the molecules to dissociate into atoms. The cloud is also too cold in most places for the hydrogen molecules to emit detectable amounts of radiation at their characteristic wavelengths. The presence of molecular hydrogen must therefore be inferred. The cloud is detected best by the emission (at 2.6 millimeters) of the next-commonest molecule, carbon monoxide, and by the emissions of still less common molecules such as formaldehyde.

One other component of a giant molecular complex is important for star formation. That component is the dust. The dust particles are sites on which the surrounding gas can collect. The dust also shields the incipient stars from ultraviolet radiation, which would disrupt the condensation. It is thought that for every 100 to 200 grams of molecular hydrogen in the giant molecular complex there is a gram of dust.

Two nearby complexes of carbon monoxide and dust (and presumably hydrogen in abundance) have been particularly well studied. They are the Orion complex (which is centered on the bright region of ionized hydrogen and newborn giant stars called the Great Nebula in Orion) and the Ophiuchus complex (which blocks the light from the center of the galaxy). In the Orion complex Becklin, Neugebauer and their associates at Cal Tech, together with Frank J.

Low of the University of Arizona, have found evidence at infrared wavelengths for the presence of condensed objects that seem to be intrinsically very red. Each one may be a newborn star still embedded in a thick cocoon of dust that the star's ultraviolet radiation has not yet blown away completely. The radiation heats the dust, which then radiates in the infrared. The star may have formed in the first place when the cooling of a small part of the complex reduced the pressure that results from the heat of a gas. The cooled region would thus have begun to collapse under the influence of its own gravitation.

In the Ophiuchus complex a group of 30 stars has been found by Gary L. Grasdalen of the University of Wyoming, Stephen E. and Karen M. Strom of the University of Arizona and Frederick J. Vrba of the U.S. Naval Observatory. The stars were not detected earlier because they are hidden by at least 30 magnitudes of optical absorption; it took an infrared search to find them. More recent observations by Charles J. Lada and Bruce A. Wilking of the University of Arizona in an extremely dense dust cloud near the star Rho Ophiuchi have revealed the presence of 20 similar stars whose optical radiation is dimmed by as much as 100 magnitudes. The stars lie only half a parsec from the ones discovered earlier.

The evidence to date suggests that giant molecular complexes give rise spontaneously to stars that have masses no greater than a few times the mass of the sun. In particular it seems they give rise to stars of the spectral classes B, A, F and G. (The sun is a star of the class G.) George H. Herbig and his associates at the Lick Observatory have found near

the borders of the well-studied giant molecular complexes some groups of small, rather dim and nebulous stars that are said to be young. Almost all of them have a brightness that varies irregularly. They are called T Tauri stars. Perhaps they are products of the interrupted condensation of nodules. They may have wandered out of the complex. Often they are seen in places where the ultraviolet radiation of newborn massive stars has blown away the dark nebula's gas and dust.

Star formation of another kind is typical, it appears, in places where an emission nebula with supergiant O and B stars embedded in it lies next to a giant molecular complex. The clearest exposition of what happens in such cases has been given by Bruce G. Elmegreen of Columbia and by Lada. According to Elmegreen and Lada, the O and B stars emit ultraviolet radiation whose pressure piles up cold gas and dust at the outer edge of the complex. The result is the condensation of protostars there. In places such as the Orion nebula the process appears to be advancing sequentially. In the Orion nebula a group of O and B supergiants is fading after a lifetime of a few tens of millions of years. The radiation from these stars has triggered the formation of a younger generation of O and B supergiants, and the younger stars in turn are now emitting radiation that eats its way slowly but persistently into the giant molecular complex, where a third set of O and B supergiants will presumably form. Why the process should give rise to O and B giants and supergiants rather than the smaller B, A, F and G stars that condense spontaneously in the complex is not yet understood.

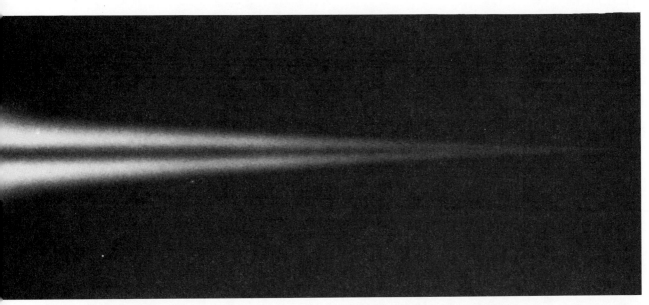

The beginning of the simulation was a mathematical description of the number of stars of a given brightness in a given direction from the solar system. The computer displayed the resulting pattern of luminosity (and a central dust lane) as it would look to an outsider.

Star Formation in Globules

I turn now to the class of dark clouds known as globules. Roughly 200 of these objects have been found within 500 parsecs of the sun. They have remarkably similar properties. Each one is dark and distinct and on a photographic plate is almost circular. No doubt they are nearly spherical. Their radius varies from .2 to .6 parsec, their mass from 20 to 200 solar masses and their internal temperature from five to 15 degrees K. They are impenetrable to visible light. On the other hand, some images recorded at near-infrared wavelengths, either photographically or by electronic imaging techniques, show the stars behind the globule. The dimming of the infrared radiation from such stars allows an estimate of the globule's content of dust.

Radio observations at 2.6 millimeters show that the globules are rich in carbon monoxide. Other molecules, notably formaldehyde and ammonia, have now been found as well. Evidently, then, a globule is a small and often isolated spherule of darkness quite similar in composition to a giant molecular complex. Presumably the globule is composed predominantly of molecular hydrogen too cold to emit detectable radiation.

The radio data also show that several of the globules are collapsing under the influence of self-gravitation. The rate of collapse is roughly half a kilometer per second, which corresponds to half a parsec per million years. Since the radius of a typical globule is half a parsec, a million years is roughly the time it takes for the collapse to be completed. The collapse again is crucial to models of star formation. In almost every model the center of the globule collapses faster than the periphery, and so a nodule forms. The collapse converts into kinetic energy the gravitational potential of the infalling matter. Eventually the energy at the center raises the temperature of the matter enough for thermonuclear fusion to begin. This signals the birth of a star. If the star is large, it emits enough radiation to blow away the gas and dust that surround it.

In short, a globule ought to give rise to a single star in about a million years. Considering the time scale of the process and estimating the number of globules in the Milky Way, one concludes that the globules might account for the formation of 25,000 stars per million years, or about a sixth of the overall rate at which the stars in the galaxy form. (The formation of stars in giant molecular complexes is perhaps a more fecund process.) In 1977, however, Richard Schwartz of the University of Missouri observed what appears to be a pair of incipient, nebulous stars that are being expelled from a very dark globule some 300 parsecs from the solar system in the part of the southern sky marked by the constellation Vela. The creation of two stars rather than one is surprising, and so is their expulsion from the globule. One wonders where the energy to expel them came from. The two stars are connected by a luminous strand that might be described somewhat fancifully as a stellar umbilical cord.

I should like to give one final example to suggest how much remains to be learned. Investigators attempting to model the processes by which stars form have tacitly assumed that the protection of a cloud of dust is required. Inside such a cloud the temperature is low and the accretion of matter is undisturbed. In particular the dust shields the interior of the cloud from disruption by ultraviolet radiation from outside sources. Consider, however, the Clouds of Magellan, which lie in the galactic corona but defy the generalization that all the content of the corona is old.

In the Large Cloud of Magellan there is a grouping of roughly 50 luminous O and B stars known as Shapley's Constellation I. It is likely that their age is no greater than 20 million years. Their velocities are on the order of only 10 kilometers per second. Hence each star has moved no more than 200 parsecs from where it was born. Within a radius of from 200 to 300 parsecs of where the stars are now, measurements at 21 centimeters have revealed five million solar masses of neutral atomic hydrogen and measurements at optical wavelengths have revealed 60,000 solar masses of ionized hydrogen. But the sky in that area is transparent. It probably harbors no giant molecular complex; it has little or no molecular hydrogen and it has no cosmic dust. How did the stars form there?

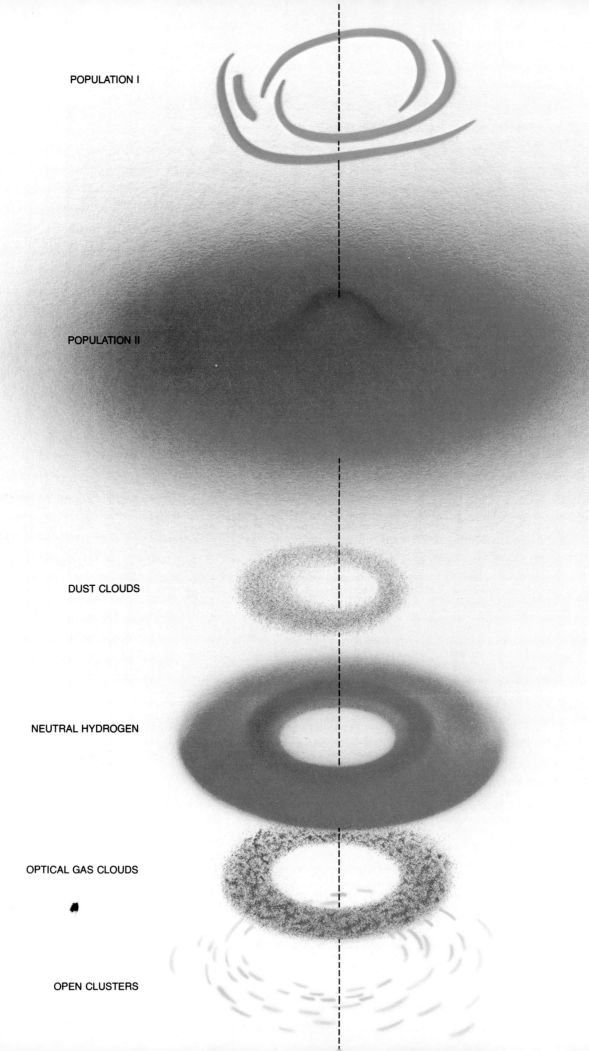

POPULATION I

POPULATION II

DUST CLOUDS

NEUTRAL HYDROGEN

OPTICAL GAS CLOUDS

OPEN CLUSTERS

The Andromeda Galaxy

by Paul W. Hodge
January, 1981

The large spiral galaxy nearest our own, it has been a laboratory for the study of the evolution of stars and galaxies. Even today it presents puzzles; for example, how are its spiral arms arrayed?

The year was 1611. Night watchmen made their rounds throughout the towns of Europe. The flames of their candles were protected from wind and rain in lanterns whose windows had a thin covering of horn. The Bavarian astronomer Simon Marius directed his telescope toward a nebulous patch of light in the region of the sky occupied by the constellation Andromeda. He likened the nebulosity to a "candle seen at night through a horn."

His description gives a good idea of how the object now called the Andromeda galaxy looks to someone with a small telescope, but it fails to suggest the place the Andromeda galaxy has held in the history of astronomy. Through large modern telescopes the Andromeda galaxy is seen to be a giant spiral galaxy. It is thought to be similar in form to our own galaxy, the Milky Way. One difference is that the Andromeda galaxy seems to be twice as large: it may include as many as 400 billion stars. At a distance of two million light-years from the solar system it is the spiral galaxy nearest our own, and it is the only giant spiral close enough for us to view it in detail. The Milky Way itself is less open to an overall inspection because, from the vantage of the earth, dust clouds hide its structure. The Andromeda galaxy has therefore been responsible for breakthroughs in the understanding of

such matters as the evolution of stars, the rotation of galaxies and the scale of distances in the universe. Sources of X rays in the galaxy and of emissions in the radio part of the electromagnetic spectrum are now being examined. With the construction of new telescopes on the earth and in orbit the Andromeda galaxy becomes a prime target for exploration in powerful new ways.

Proof of Distance

The first modern study of the Andromeda galaxy was made almost 100 years ago, when photography first presented a way to record light too dim for the eye to see, and thereby to probe deep into space. With a 20-inch telescope Isaac Roberts took the first photographs of the Andromeda galaxy that showed its spiral structure. The photographs also suggested the presence of faint stars in the outer parts of the spiral, but that clue to the nature of the object was not understood at the time. Instead the Great Nebula in Andromeda was taken to be a cloud of gas that might eventually condense to form a star with a planetary system. It seemed to be the largest, the brightest and therefore probably the nearest among hundreds of similar nebulas. It was thus thought to be relatively close to the solar system.

The idea that star systems lay beyond

the Milky Way soon occurred to several investigators, including Edwin P. Hubble of the Mount Wilson Observatory. In 1925 Hubble demonstrated that the small, inconspicuous nebula NGC 6822 is a distant aggregation of stars. Meanwhile the great spiral in Andromeda occupied much of his attention. It was the subject of his landmark paper, "A Spiral Nebula as a Stellar System," published in 1929. Hubble's many photographs of the object showed a huge, amorphous bulge of light surrounded by tightly wound spiral arms consisting of dust clouds, star clusters and thousands of points of light, each one a star.

Hubble's proof of the great distance of the Andromeda nebula derived from his discovery of 40 pulsating stars in the spiral. The examination of successive photographs of the galaxy showed that the stars were periodically brightening and dimming. Hubble recognized the stars as Cepheid variables, which are also found in the Milky Way. Harlow Shapley had already shown that Cepheid variables could serve as astronomical yardsticks: the intrinsic luminosity of a Cepheid variable is proportional to the period of its brightening and dimming, whereas its apparent luminosity can of course be measured directly. The ratio of the two luminosities is proportional to the square of the distance of the star. Using Shapley's calibration of the relation between period and intrinsic luminosity, Hubble concluded that the Andromeda spiral must be almost a million light-years from the earth, far beyond the edge of the Milky Way. Subsequent work has shown that the Andromeda galaxy is actually more than twice as far as Hubble calculated. I shall return to this matter below.

By comparing a succession of photographs Hubble also discovered 63 stars that flared up in luminosity and then slowly dimmed. These are novas. It is now thought the flares of light are emitted when gas escapes from a giant star, falls to the surface of a hot, dense companion star and explodes in a sudden episode of nuclear fusion. In effect the

EXPLODED DIAGRAM shows the distribution of the principal types of object in the Andromeda galaxy. For each tier of the illustration the vertical broken line marks the center of the galaxy and the horizontal direction corresponds to the major axis of the galaxy's disklike image in the sky. Stars of Population I (*top tier*) are young and blue. They are arrayed in segments that suggest a highly imperfect spiral pattern. The pattern in the illustration reflects the author's interpretation of the distribution of the stars. Stars of Population II (*second tier*) are old and red. They make up the central bulge of the galaxy and to a lesser extent the galactic disk. They extend in the illustration to a radius of some 130,000 light-years, which places them beyond the optical image of the galaxy in the picture on the next page. Stars of Population II also make up the aggregations known as globular clusters (not shown), which are scattered above and below the plane of the galaxy. Dust clouds (*third tier*) are visible because they redden or even block the light from the stars behind them. Their distribution is greatest toward the northeast (the left in the illustration). Neutral (un-ionized) hydrogen (*fourth tier*) is detected by its emission in the radio part of the electromagnetic spectrum. Its distribution has the shape of a doughnut. Evidence of star formation is provided by optical gas clouds (*fifth tier*), which are made luminous by their proximity to hot, young stars. Their distribution is more or less coextensive with the region of spiral-arm segments. Open clusters (*bottom tier*) are loose aggregations of stars of Population I. Their distribution, however, does not evoke a clear spiral pattern.

surface of the companion star becomes a hydrogen bomb. The novas in the Milky Way show that at maximum brightness these objects are approximately equal in luminosity to the brightest normal stars. In the Andromeda galaxy Hubble found that the novas were quite faint but were nonetheless similar in luminosity to the brightest stars in that galaxy. This was further proof of the galaxy's great distance.

In 1885 another event identified as a nova had been observed near the center of the Andromeda nebula. At its brightest it was extraordinarily luminous. Indeed, it was almost visible to the unaided eye. In later debates the event, called S Andromeda, was cited as evidence that the nebula must be within our galaxy. Otherwise, it was argued, S An-

dromeda would have been much too bright to be a nova. Hubble's study of the many novas in the Andromeda galaxy convinced him that S Andromeda belonged to an exceptional class of stellar explosions just being recognized at the time. They are now called supernovas. Quite unlike ordinary novas, supernovas are explosions that destroy a star. So far S Andromeda is the only supernova observed in the Andromeda galaxy. Statistics in other galaxies suggest that we should not be surprised to see another one any day.

Baade's Discoveries

The next important figure in the exploration of the Andromeda galaxy was Walter Baade, who systematically ex-

plored the galaxy with the 100-inch telescope on Mount Wilson. His work led to two discoveries that thoroughly revised our understanding of stars and our scale of distances in the universe.

The first discovery was made in 1944. Baade was a German national at the time, and so he was not able to join the American war effort. This left him almost the sole user of the 100-inch telescope and allowed him ample time for experimenting with new photographic emulsions and with optical filters of various colors. During the war the nearby city of Los Angeles held occasional defensive blackouts. On those nights the sky was unusually dark, and Baade was able to extend his photography of the Andromeda galaxy to record progressively fainter features.

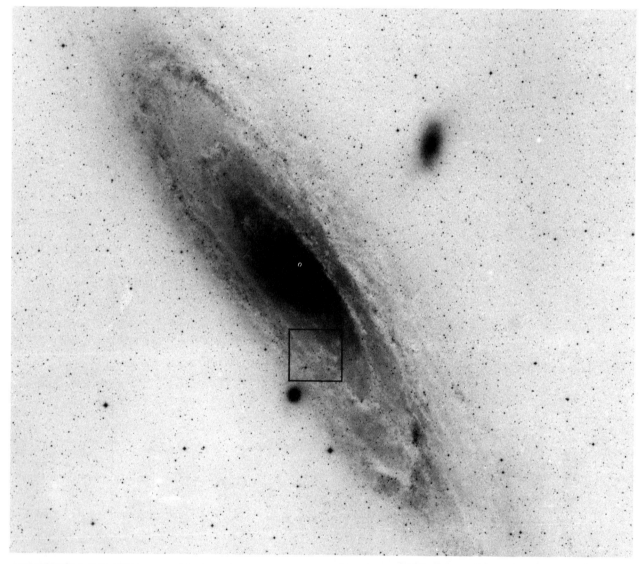

ANDROMEDA GALAXY is arrayed from northeast to southwest in the sky above the Northern Hemisphere. Its plane is oblique with respect to the earth, so that the nearest part of the galaxy is the northern edge of its disk. The photograph shown here is printed as a negative; hence the dark parts of the galaxy consist of stars. The white parts consist of dust, which conceals the stars behind it. A prominent dust lane lies to the right of the galaxy's central bulge. The dark spot below the bulge is M32, a companion galaxy. A second companion galaxy, NGC 205, lies above and to the right of the bulge. Two other companion galaxies are outside the field of view. The diameter of this image of the Andromeda galaxy is about 125,000 light-years, and the galaxy is two million light-years from the earth. The galaxy thus subtends an angle of three degrees in the sky, or six times the angle subtended by the moon. The black rectangle marks the part of the galaxy in the illustration on the opposite page. The photograph was made with the 48-inch Schmidt telescope on Palomar Mountain.

Baade was trying to understand why the spiral arms of the Andromeda galaxy were well resolved into individual stars in photographs of the galaxy, whereas the bright, amorphous central region defied resolution even on Hubble's best photographic plates. The light from the central region doubtless came from millions of stars. Why could none of them be seen? To compound the mystery, four small, somewhat featureless galaxies are companions of the giant spiral (two are closer and two are more distant), and these too could not be resolved into stars.

Eventually Baade tried a fortunate combination: photographic emulsions sensitive to red light, a red filter, perfect atmospheric conditions, a blacked-out Los Angeles and extremely long exposure times. The resulting photographs not only resolved the stars in the central bulge of the Andromeda galaxy and in the four companion galaxies; they also led Baade to distinguish two stellar populations. The elusive stars that appeared on Baade's plates were red-giant stars, large but nonetheless too faint to be resolved by the emulsions previously employed. They belonged to a class of stars that Baade called Population II. Such stars were the same, Baade contended, as the ones that make up the aggregations in the Milky Way called globular clusters.

In number the stars of Population II predominate at the center of the Andromeda galaxy and in that galaxy's globular clusters, which are scattered throughout a spherical volume extending above and below the plane of the galactic disk. In mass (although not in luminosity) they probably predominate throughout the Andromeda galaxy. Subsequent work by Baade, his students and others showed that Population II stars are all about 12 billion years old, which makes them almost as old as the universe. In contrast, Population I includes the bright blue stars that make up the spiral arms of the Andromeda galaxy. These stars are thought to be young. Population I also includes the gas and dust that tend to enmesh bright blue stars.

Baade's second major discovery was also largely dependent on the Andromeda galaxy. The globular clusters of the Milky Way include a few Cepheid variables, and Shapley had used them to calibrate the relation of period to luminosity for the Cepheids generally. Baade noticed, however, that in the globular clusters of the Milky Way the brightest red-giant stars had about the same apparent luminosity as the Cepheid variables whose periods were from 30 to 40 days. In the Andromeda galaxy the brightest red giants were fainter than the Cepheids with those periods that did not lie in globular clusters.

The resolution of this discrepancy was announced at the 1952 meeting of

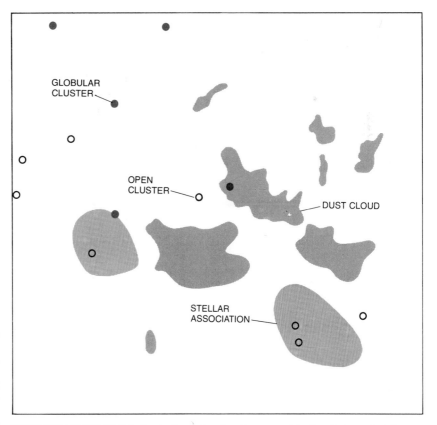

WEALTH OF OBJECTS in the Andromeda galaxy is suggested by the photograph at the top, which shows a small part of the galaxy. The map at the bottom identifies some of the objects. Both "open clusters" and "globular clusters" are aggregations of stars. In the Milky Way galaxy the former are younger and less compact than the latter, but that distinction is less clear for the clusters in the Andromeda galaxy. Each cluster in the Andromeda galaxy is apparent as simply a patch of light. "Stellar associations" are loose aggregations of very young stars. The photograph was made with the four-meter telescope at Kitt Peak National Observatory.

In the map: GLOBULAR CLUSTER, OPEN CLUSTER, DUST CLOUD, STELLAR ASSOCIATION the International Astronomical Union in Rome, when Baade showed that the Cepheid variables in the globular clusters of the Milky Way are a special type of star found only in the company of stars of Population II. They are not the same as the Cepheid variables in the spiral arms of the Andromeda galaxy or in those of the Milky Way. The Cepheids in globular clusters are intrinsically about four times fainter than spiral-arm Cepheids with the same period of variation. Since the apparent brightness of an object varies as the square of the distance, it follows that the Cepheids studied by Hubble in the Andromeda galaxy must be twice as far away as Hubble had calculated. The putative distances of all the galaxies depended on Hubble's chain of reasoning, and so Baade's second discovery doubled the size of the universe.

Star Formation

Baade's interest in stellar populations led him to note that the gas and dust in the Andromeda galaxy seemed to be associated almost exclusively with Population I stars. This association was circumstantial evidence that the gas and dust mark areas where new stars are forming. In brief, the bright blue stars of Population I are young because a star as luminous as they are cannot have maintained that output of energy for long. Evidently, therefore, the gas and dust provide the raw material from which the stars are formed. The dust particles could be sites of accretion for the gas.

In trying to discover both how stars form and how the regions of star formation are arrayed in a spiral galaxy, Baade examined the distribution of hot gas clouds. Such clouds are illuminated by the bright, young stars embedded in them, but they absorb the light and then emit it only at certain wavelengths, which correspond in energy to excitations of the electrons in the atoms of the gas. By making photographs with filters that allowed only light of those wavelengths to pass, Baade was able to map the locations of 688 gas clouds in the

Andromeda galaxy. He found that they were concentrated in the spiral arms. They were most conspicuous at the middle distance along the arms, some 30,000 to 40,000 light-years from the center of the galaxy. That is where the bright blue stars are present in the greatest number, and so it is also in such regions that the newest stars in the galaxy are now forming.

With the development of radio astronomy new methods became available for examining the formation of stars. The most important single advance was the discovery of the radio emission of neutral (that is, un-ionized) hydrogen gas at a wavelength of 21 centimeters. Each photon, or quantum, of the emission arises when the intrinsic angular momentum of the electron in a hydrogen atom inverts with respect to that of the proton in the atom. It is also necessary that the atom be in its lowest energy state, and so the hydrogen atoms emitting the radiation cannot often be in collision. (The collisions would excite the hydrogen atoms to higher energy levels.) Hence the astronomical sources of the 21-centimeter radiation are cold, thin clouds of gas.

The landmark study of neutral hydrogen came in 1966, when Morton S. Roberts of the National Radio Astronomy Observatory employed the 300-foot radio telescope at Green Bank, W.Va., to amass the data for a high-resolution map of the Andromeda galaxy's radiation at 21 centimeters. Instead of a disk of gas coincident with the galactic distribution of stars, Roberts found that the distribution of neutral hydrogen resembled a giant doughnut, with a hole in the middle and a maximum density about 40,000 light-years from the center, a distance that corresponds to both the brightest parts of the spiral arms and the maximum concentration of the gas clouds mapped by Baade. To this extent the distributions of hot and cold hydrogen match.

Beyond the thickest part of the doughnut the two distributions differ. The hot gas clouds taper off at about 50,000 light-years. The cold neutral hydrogen could be detected out to twice that distance. This kind of pattern is now known to be fairly common in giant spiral galaxies, and it is understood to result from a mixture of causes. In the inner region of such a galaxy stars of Population II predominate. Star formation was complete long ago, and little gas is left. In the middle region, where gas is still abundant, passing shock waves in the density of the gas evidently trigger the condensation of new stars. In the outer region such shock waves are weak or absent, and so the density remains too low for new stars to condense.

In recent years still more has been learned about the distribution of neutral hydrogen, as a result of the development of radio telescopes in which signals

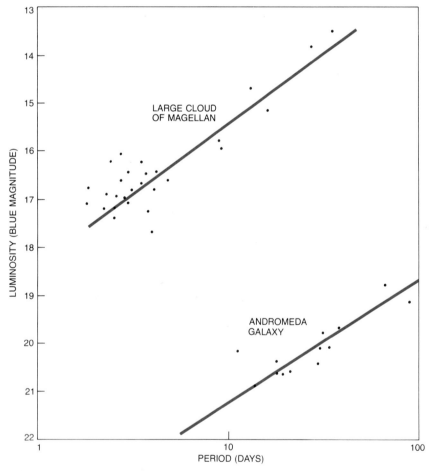

DIFFERENCE IN LUMINOSITY between the Cepheid variable stars in the Andromeda galaxy and the ones in the Large Cloud of Magellan (a small companion galaxy of the Milky Way galaxy) allows a calculation of the distance to the Andromeda galaxy. The luminosity of each such star in the Large Cloud of Magellan varies with a periodicity (*horizontal axis*) that is proportional to its brightness (*vertical axis*). The Cepheids in the Andromeda galaxy exhibit the same behavior, but the stars appear to be fainter as a result of their greater distance. In particular, a difference of five magnitudes indicates a hundredfold difference in brightness and a tenfold difference in distance. The Large Cloud of Magellan is 200,000 light-years from earth.

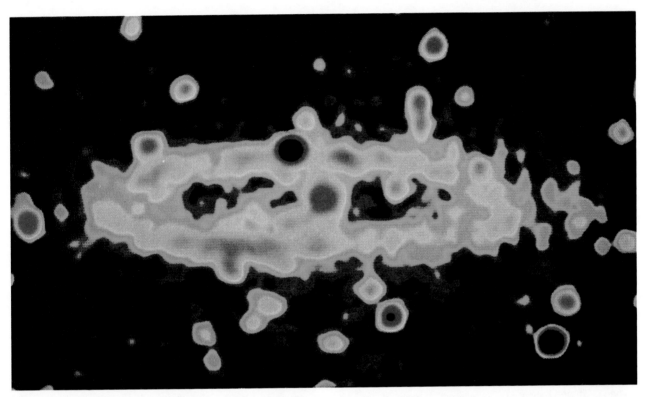

RADIO EMISSION of the Andromeda galaxy at a wavelength of 11 centimeters is displayed by a computer at the Rheinisches Landesmuseum at Bonn in West Germany. Red signifies the highest intensity and violet the lowest. Thus the strongest source of emission is at the center of the galaxy, a location that probably is notable for the concentration of the remnants of exploded stars. The overall symmetry of the pattern is produced by sources at a radius of about 30,000 light-years from the galactic center. That distance also characterizes the distribution of hot gas clouds, which mark regions where stars form. The data were obtained at the 100-meter radio telescope at Effelsberg in West Germany by Rainer Beck, Elly Berkhuijsen and Richard Wielebinski of Max Planck Institute for Radio Astronomy.

from several parabolic antennas are processed to yield a resolution equal to that of a telescope the size of the distance of their separation. The first studies to be completed with this new generation of instruments include the work of D. T. Emerson and his colleagues at the University of Cambridge, who employed the Cambridge Half-Mile Telescope to show that the hole in the doughnut of neutral hydrogen in the Andromeda galaxy extends to a distance of about 12,000 light-years from the galactic center. From that distance outward the distributions of hydrogen and dust seem to match closely. On the other hand, the Cambridge investigators found hydrogen some 105,000 light-years from the center of the Andromeda galaxy along the southwestern major axis of the galaxy. (The galaxy slants from northeast to southwest in the sky.) That puts it farther from the center than any part of the galaxy's image in an optical telescope.

Newest Hydrogen Maps

An even more detailed study of neutral hydrogen in the Andromeda galaxy has now been made by Estaban Bajaja of the Westerbork Observatory, who employed the Westerbork Aperture Synthesis Telescope in the Netherlands. Although not all the results have yet been published, Bajaja has found a close correspondence between hydrogen and optically visible dust only along the northeastern major axis and therefore in only half the galaxy. On the other side of the galactic center the arrangement of both gas and dust is poorly defined.

Bajaja has also investigated the motion of the neutral hydrogen. He finds that it is deviating from a strictly circular trajectory. The simplest hypothesis about a spiral galaxy's motion is that all its components rotate about the massive galactic center. The stars of Population II are thought to have formed before much of the galaxy had condensed into a disk. Their orbits are highly elliptical. In contrast the orbits of young stars and of gas and dust in the disk of Population I are thought to be almost circular, like the orbits of planets around the sun.

The new discoveries clearly show that this assumption is sometimes quite wrong. Of the three apparent arms of hydrogen in the northeast half of the Andromeda galaxy, parts of the innermost arm are plummeting inward toward the center of the galaxy with a velocity of at least 100 kilometers per second. This is in addition to their motion around the center. The cause of the inward velocity is a mystery. The next arm out does not show it. Is the noncircular motion related to the complex dynamics of a warping of the galactic disk? Is it caused by the companion galaxies, or has some explosive event in the past disrupted this part of the galaxy? The answer probably lies in the increasingly detailed data the radio-telescope arrays are now gathering.

While the instrument at Westerbork has been mapping the radio emission of neutral hydrogen at 21 centimeters, a 100-meter single-dish radio telescope near Bonn in West Germany has enabled Elly Berkhuijsen and her colleagues at the Max Planck Institute for Radio Astronomy to map the emission at 11 centimeters and deduce the nature of its sources. Earlier G. G. Pooley of Cambridge had shown that most of the Andromeda galaxy's 11-centimeter radio emission came from the center of the galaxy and also from the parts of the spiral arms that are bright in optical images.

Piet van der Kruit of the Leiden Observatory and Yervant Terzian and Bruce Balick and their co-workers at Cornell University confirmed Pooley's results at other wavelengths. They also demonstrated from the relative intensities of the radiation at those wavelengths that a large proportion of the radiation could not be thermal. A source of thermal radiation, such as a cloud of hot gas, emits radio noise whose spectrum has a characteristic rise and fall determined by the cloud's size and temperature. The intensities recorded by the Cornell investigators do not conform to that shape. Instead much of the radiation must be generated by some nonthermal mechanism. It might, for example, be synchrotron radiation, which is given off by electrons moving at nearly the speed of light through a magnetic field. Synchrotron radiation includes electromagnetic waves with almost equal intensity from X rays to long radio waves. It arises in places where large amounts of energy are released, such as the vicinity of supernova explosions and the collapsed supermassive objects (neutron stars or black holes) the explosions are thought to leave behind.

Berkhuijsen's maps showed that thermal radiation comes mostly from the vicinity of the doughnut of neutral hydrogen, about 30,000 light-years from the center of the Andromeda galaxy. It is therefore probably given off by the numerous hot gas clouds that are commonest there. The nonthermal radiation is more broadly based. It has a sharply peaked source at the center of the galaxy, but it extends more or less evenly to a distance from the center of from 40,000 to 50,000 light-years. Supernova remnants in the Milky Way galaxy (and the seven remnants discovered so far in the Andromeda galaxy by Vera C. Rubin, Cidambi K. Kumar and W. Kent Ford, Jr., of the Department of Terrestrial Magnetism of the Carnegie Institution of Washington) have a similar distribution. Thus the nonthermal radiation may be explained largely as the combined noise from the remnants of supernovas, perhaps including the spectacular S Andromeda of 1885.

Other Objects

A kind of bright gas cloud that resembles a smoke ring is thought to be the shed outer atmosphere of an aging or dying star. The ring is called a planetary nebula because although it is nebulous, its image resembles that of a planet in that the image is compact and disklike. Although Baade found five such objects in the Andromeda galaxy, they have been discovered there in large numbers only in the past two or three years. Holland Ford and George Jacoby of the University of California at Los Angeles have exploited new techniques of filtering and image intensification to record images that would have been too faint for Baade to record. With the 120-inch telescope of the Lick Observatory they found 315 planetary nebulas. They calculate that the total number in the Andromeda galaxy is roughly 10,000. Since planetary nebulas mark dying stars, such nebulas map the galaxy's decay. In the bulge at the center of the galaxy about five stars per century are apparently dying. Over the past few billion years the gas released in this way

should have accumulated in a gaseous disk that rotates about the bulge. Radio astronomers have found a central disk of gas whose mass is just about the expected one.

A similar application of modern image intensifiers and filters has led to progress in the exploration of the hot gas clouds in the Andromeda galaxy. A survey much superior to Baade's effort of the 1950's was recently completed by a French group led by G. Courtés at the two-meter telescope of the Haute-Provence Observatory. The measured velocities of the clouds, combined with the velocities of neutral hydrogen, now provide a clear picture of the rotation of the galaxy. Moreover, the velocity of an object in orbit depends on the amount of mass inside the orbit as well as on the mass near the orbiting object, so that from the orbital speeds of objects in the Andromeda galaxy it is possible to deduce the distribution of mass in the galaxy. The most recent attempts to deduce the total mass yield values between 200 billion and 400 billion times the mass of the sun, or as much as twice the mass of the Milky Way galaxy. Even so, the orbital velocity of hydrogen in the outer parts of the Andromeda galaxy hints

that large amounts of unseen matter may form a huge halo.

One way to test the current value of the mass of the Andromeda galaxy would be to measure the gravitational interaction of the galaxy's mass with some other objects, preferably objects at great distances from the galactic center. Realizing that the globular clusters in the galaxy's Population II might be the needed objects, F. D. A. Hartwick of the University of Victoria, Sidney van den Bergh of the Herzberg Institute of Astronomy in Victoria and Wallace L. W. Sargent of the California Institute of Technology recently collaborated on a search for the globular clusters on photographic plates made with the new four-meter telescope of the Kitt Peak National Observatory in Arizona. They have catalogued 355 probable globular clusters, doubling the number that had previously been discovered by astronomers beginning with Hubble. The 355 globular clusters are almost three times as many as are known to lie in Population II of the Milky Way galaxy. The abundance of globular clusters is evidence in itself that the mass of the Andromeda galaxy is great. The orbits of the clusters have not yet been deter-

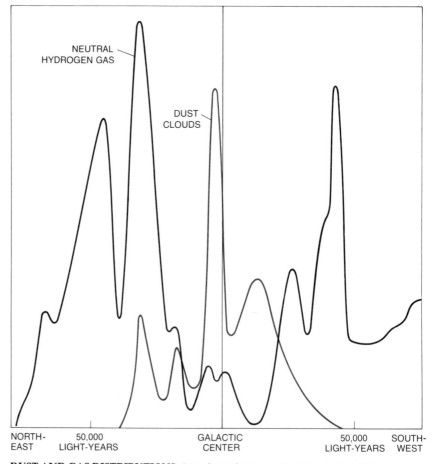

NEUTRAL
HYDROGEN GAS

DUST
CLOUDS

NORTH-EAST   50,000 LIGHT-YEARS   GALACTIC CENTER   50,000 LIGHT-YEARS   SOUTH-WEST

DUST AND GAS DISTRIBUTIONS along the major axis of the Andromeda galaxy are compared. The dust particles provide sites on which the gas can accrete, and so it is thought that both types of matter are required for the formation of stars. The distributions agree moderately well toward the northeast side of the galactic disk but not at all toward the southwest side.

mined, however, and so the question of the total mass remains open.

It is known from a study of the spectra and the colors of the brighter globular clusters by van den Bergh that the early history of the Andromeda galaxy must have been puzzlingly different from the history of the Milky Way galaxy. In the Milky Way galaxy the globular clusters in the outer regions are very poor in heavy elements. The sun is much richer. This difference is usually interpreted as part of a general pattern. The ancient stars of Population II are almost purely hydrogen and helium, the elements created in abundance in the early universe, whereas the younger disk of Population I stars includes debris from a multitude of dying stars, which synthesized heavy elements by thermonuclear fusion. Surprisingly, the spectra of the globular clusters in the Andromeda galaxy reveal a diversity of patterns of heavy-element abundance, and there is no relation at all between the abundance of heavy elements in a cluster and the position of the cluster in the galaxy.

The Spiral-Arm Controversy

Clusters of another kind have also been found in the Andromeda galaxy. They are open clusters, or loose aggregations of stars. Open clusters are younger than globular clusters, and they lie in the plane of the galaxy, along with the rest of Population I. Except for a couple of examples Hubble identified they had not been found in the Andromeda galaxy. Last year, however, I took advantage of the wide field of view of the new four-meter telescope at the Kitt Peak National Observatory to survey the galaxy and identify 403 of them.

Most of the 403 are only 60 light-years or so across. They have nonetheless allowed me to begin to trace the recent history of star formation in the Andromeda galaxy. The crucial point is the likelihood that the stars of an open cluster all form simultaneously. On the other hand, the distribution of various types of stars in the cluster changes over a period of time. Thus a particular set of statistics implies a particular age. In this way I have shown that the rate of star formation has varied across the disk of the Andromeda galaxy. Recently it has been anomalously great about 30,000 light-years from the center, where neutral hydrogen and bright stars are concentrated.

The open clusters are now being invoked in the controversy over the configuration of the spiral arms of the Andromeda galaxy. The first attempt to delineate the arms was that of Halton C. Arp of the Hale Observatories, who based his work on the distribution of gas clouds in the galaxy. Arp found that a two-armed spiral with the arms trailing the direction in which the galaxy rotates fitted the scatter of data

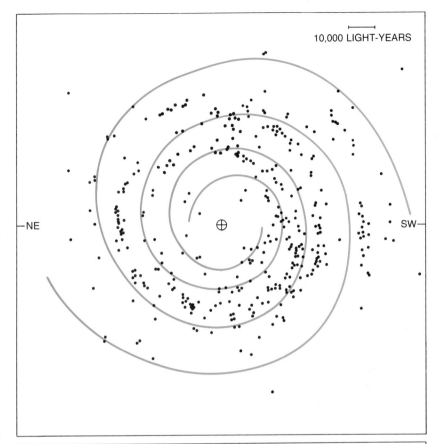

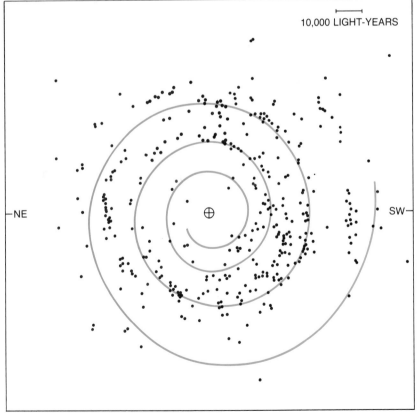

SPIRAL-ARM CONTROVERSY arose when Halton C. Arp of the Hale Observatories saw in the Andromeda galaxy a two-armed spiral form (*top*), whereas Agris Kalnajs of the Mount Stromlo Observatory in Australia discerned a one-armed pattern (*bottom*), with the arm spreading outward in the opposite direction. (The rotation of the galaxy is counterclockwise when the galactic disk is viewed, as it is here, from below.) The dots in the illustration mark positions of open clusters, which consist of young stars. It is young stars that form the spiral-arm segments.

points fairly well. He suggested that the imperfections of the fit might be due to the gravitational disturbance of the galactic disk by the mass of M32, one of the four companions of the Andromeda galaxy.

A recent analysis done with the aid of a computer by Gene G. Byrd of the University of Alabama seems to confirm this and at the same time seems to account for the infalling observed in the distribution of neutral hydrogen. On the other hand, Agris Kalnajs of the Mount Stromlo Observatory in Australia has concluded from the distribution of clouds of hydrogen that the galaxy has only one spiral arm and that it spirals in the direction opposite to that of the two arms plotted by Arp. It is thus a leading arm: its free end points in the direction of the galaxy's rotation.

There is no unambiguous instance of a leading one-armed spiral galaxy anywhere else in the universe. Kalnajs has noted, however, that the presence of M32 may be the explanation. If the period of rotation of the Andromeda galaxy and the period of rotation of M32 about the Andromeda galaxy have a common divisor, a gravitational resonance between the two galaxies could give rise to the leading arm. Recently a group of French, Swiss and Greek astronomers examined a large collection of positional data for Population I objects and attempted to discern a spiral structure. They too concluded that the best fit was that of a leading one-armed spiral.

The controversy remains unresolved. The distribution of gas clouds seems to fit a one-armed pattern fairly well. It turns out, however, that the distribution of open clusters fits only a two-armed spiral that is gravitationally warped. Moreover, the dust clouds fit no spiral pattern well. The discrepancies show how imperfect most galactic spiral patterns are, and also how imperfect our understanding of galactic spirals is.

The Andromeda galaxy continues to be the subject both of observations made possible by new techniques and of searches for instances of newly discovered types of object. Maps now plot the distribution of carbon monoxide in the Andromeda galaxy; they show that it is closely correlated with the neutral hydrogen. Moreover, just last year the Einstein Orbiting X-Ray Observatory detected 69 X-ray sources in the Andromeda galaxy. Astronomers at the Center for Astrophysics of the Harvard College Observatory and at the Smithsonian Astrophysical Observatory think some of the sources are globular clusters, some are concentrations of Population I objects and some are supernova remnants. The precise nature of the high-energy processes that give rise to the radiation may soon be deduced. The Andromeda galaxy has provided astronomers with a rich lode of information. It will no doubt reveal much more.

Dark Matter in Spiral Galaxies

by Vera C. Rubin
June, 1983

*It appears that much of the matter in spiral galaxies
emits no light. Moreover, it is not concentrated
near the center of the galaxies*

After evidence was obtained (in the 1920's) that the universe is expanding it became reasonable to ask: Will the universe continue to expand indefinitely or is there enough mass in it for the mutual attraction of its constituents to retard the expansion and finally bring it to a halt? Most cosmologists agree that the universe started in a big bang 10 to 20 billion years ago from an infinitely small and dense state and that it has been expanding ever since. It can be calculated that the critical density of matter needed to brake the expansion and "close" the universe is on the order of 5×10^{-30} gram per cubic centimeter, which is equal to about three hydrogen atoms per cubic meter. The amount of luminous matter in the form of galaxies, however, comes to only about 7.5×10^{-32} gram per cubic centimeter. Therefore if the expansion of the universe is to stop, the density of the invisible matter must exceed the density of the luminous matter by a factor of roughly 70.

With this factor in mind astronomers over the past half century have sought to determine the mass of the galaxies that populate the universe out to the limits of observation. From the luminosity of typical galaxies one can estimate that they have a mass ranging from a few billion to a few trillion times the mass of the sun. The actual stellar population of a galaxy is of course highly diverse. Some stars are 10,000 times more luminous than the sun per unit of mass; others are only a small fraction as luminous. Given this diversity one would like to know: Is the distribution of luminosity in galaxies a reliable indicator of the distribution of mass? And, by extrapolation, is the distribution of luminosity in galaxies a reliable indicator of the distribution of mass in the universe?

My colleagues and I in the Department of Terrestrial Magnetism of the Carnegie Institution of Washington have sought to answer these questions by measuring the rotational velocity of selected galaxies at various distances from their center of rotation. It has been known for a long time that outside the bright nucleus of a typical spiral galaxy the luminosity of the galaxy falls off rapidly with distance from the center. If luminosity were a true indicator of mass, most of the mass would be concentrated toward the center. Outside the nucleus the rotational velocity would fall off inversely as the square root of the distance, in conformity with Kepler's law for the orbital velocity of bodies in the solar system. Instead it has been found that the rotational velocity of spiral galaxies in a diverse sample either remains constant with increasing distance from the center or rises slightly out as far as it is possible to make measurements. This unexpected result indicates that the falloff in luminous mass with distance from the center is balanced by an increase in nonluminous mass.

Our results, taken together with those of many other workers who have attacked the mass question in other ways, now make it possible to say with some confidence that the distribution of light is not a valid indicator of the distribution of mass either in galaxies or in the universe at large. As much as 90 percent of the mass of the universe is evidently not radiating at any wavelength with enough intensity to be detected on the earth. Originally astronomers described the nonluminous component as "missing matter." Today they recognize that it is not missing; it is just not visible. Such dark matter could be in the form of extremely dim stars of low mass, of large planets like Jupiter or of black holes, either small or massive. Other candidates include neutrinos (if indeed they have mass, as recent work suggests) or such hypothetical particles as magnetic monopoles or gravitinos.

Early in this century it was reasona-

ble for astronomers to assume that the distribution of luminous matter, wherever it was found, coincided with the distribution of mass. Nearly 50 years ago, however, Sinclair Smith and Fritz Zwicky of the California Institute of Technology discovered that in some large clusters of galaxies the individual members are moving so rapidly that their mutual gravitational attraction is insufficient to keep the clusters from flying apart. Either such clusters should be dissolving or there must be enough dark matter present to hold them together. Almost all the evidence suggests that clusters of galaxies are stable configurations. Hence the early observations of Smith and Zwicky marshaled the first evidence that such clusters harbor matter both luminous and nonluminous.

Recent work by many other astronomers has strengthened this conclusion. Studies of the dynamics of individual galaxies, including our own galaxy, of pairs of galaxies, of groups of galaxies and of clusters of galaxies all point to a component of unobservable but ubiquitous mass. Such studies detect the presence of nonluminous mass solely by its gravitational effects.

For the past several years W. Kent Ford, Jr., Norbert Thonnard, David Burstein and I have sought to learn about the distribution of mass in the universe by investigating the distribution of matter within galaxies with a structure similar to that of our own galaxy, namely the general class of spiral galaxies. We have adopted this approach because spiral galaxies have a geometry favorable for the identification of mass, whether it is luminous or nonluminous, and modern large telescopes equipped with image-tube spectrographs make it possible to complete an observation of a single galaxy with an exposure of about three hours. Before I describe our observations it will be helpful if I review how celestial objects respond to the gravitational force acting on them and how that response can reveal the large-scale distribution of matter.

Toward the end of the 17th century Robert Hooke suspected that the planets were subject to a gravitational force from the sun whose intensity decreased inversely as the square of the distance. Isaac Newton then recognized that all pairs of objects in the universe have a gravitational attraction for each other that is proportional to the product of their masses and inversely proportional to the square of the distance between them. In other words, if the distance between the objects is increased by, say, a factor of two, their mutual attraction decreases by a factor of four.

For planets in orbit around the sun, which embodies essentially all the mass in the solar system, the decrease in gravitational attraction with distance is exactly paralleled by a decrease in the velocity needed to hold the planet in its orbit. Therefore Mercury, lying at .39 astronomical unit from the sun (that is, .39 of the mean distance between the sun and the earth), has an orbital velocity of about 47.9 kilometers per second. Pluto, 100 times farther away at a mean distance of 39.5 astronomical units, has an orbital velocity only a tenth that of Mercury, or 4.7 kilometers per second. Spiral galaxies rotate because they retain the angular momentum and the orbital momentum of the initial clumps of gas from which they formed.

In a spiral galaxy the gas, dust and stars in the disk of the galaxy (together with any associated planets and their satellites) are all in orbit around a common center. Like the planets in the solar system, the gas and stars move in response to the combined gravitational attraction of all the other mass. If the galaxy is visualized as a spheroid, the gravitational attraction due to the mass M_r lying between the center and an object of mass m in an equatorial orbit at a distance r from the center is given by Newton's law GmM_r/r^2, where G is the constant of gravitation. If the galaxy is neither contracting nor expanding, the gravitational force is exactly equal to the centrifugal force on the mass at distance r: $GmM_r/r^2 = mV_r^2/r$, where V_r is the orbital velocity.

When this equation is solved for V_r, the value of m drops out and the velocity of a body at distance r from the center is determined only by the mass M_r inward from its position. If, as in the solar system, virtually all the mass is near the center, then the velocities outward from the center decrease as $1/r^2$. Such a decrease in orbital velocity is called Keplerian after Johannes Kepler, who first stated the laws of planetary motion.

In a galaxy the brightness is strongly peaked near the center and falls off rapidly with distance. Astronomers had long assumed that the mass too decreased rapidly with distance, in accordance with the distribution of luminosity. Hence it was expected that stars at increasing distances from the center would have decreasing Keplerian orbital velocities. Until recently few velocity observations had been made in the faint outer regions of galaxies, either to confirm this expectation or refute it.

Although the forms of spiral galaxies are exceedingly diverse, astronomers are able to group them into three useful classes following a scheme proposed some 60 years ago by Edwin P. Hubble. Galaxies designated Sa have a large central bulge surrounded by tightly wound smooth arms in which "knots," or bright regions, are barely resolved. Sb galaxies have a less pronounced central bulge and more open arms with more pronounced knots. Sc galaxies have a small central bulge and well-separated arms speckled with distinct luminous segments. The progression from type Sa to type Sc is one of decreasing prominence of the central bulge and increasing prominence of the disk rotating about it. That the disk is indeed rotating is assumed on simple dynamical grounds.

Within each type there are systematic variations in size and luminosity. For example, Sc galaxies range from small, low-luminosity, low-mass objects to galaxies of enormous luminosity and mass. For completeness, therefore, the study of the dynamics of galaxies should include not only objects with a range of morphological types but also objects with a range of luminosities.

Only for the closest stars in our own galaxy is it possible to detect motion by observing the changing position of the star against the background of more distant stars and galaxies on the celestial sphere. Even for the Andromeda galaxy, the large spiral galaxy closest to our own, it would take some 20,000 years for an orbital velocity of 200 kilometers per second (a velocity comparable to the sun's) to carry a star one second of arc across the sky. This is the minimum angular separation that can be detected optically from the earth. To study the motions in galaxies a different method is needed, one based on the phenomenon of the Doppler shift.

Doppler shifts are shifts in the frequency of waves from a source caused by the motion of the source toward or away from the observer. When the spectrum of the bright nucleus of a spiral galaxy is recorded, the absorption lines arising from the constituent stars are shifted toward the long-wavelength (red) end of the spectrum compared with the same lines in spectra made in laboratories on the earth. Such red shifts in the spectra of all but a few of the nearest galaxies, first observed in about 1915 by V. M. Slipher of the Lowell Observatory, provide the evidence that

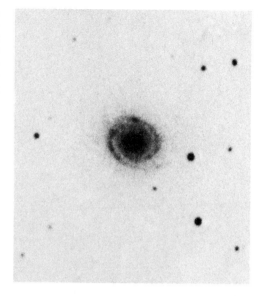

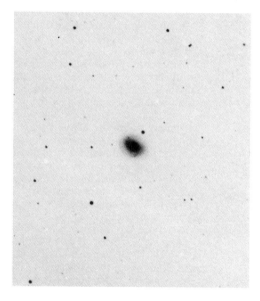

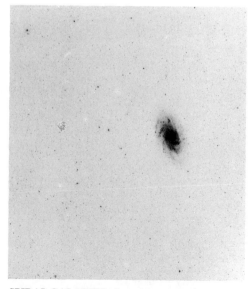

SPIRAL GALAXIES whose unseen mass has been investigated by the author fall into three principal classes: Sa, Sb and Sc. Within each

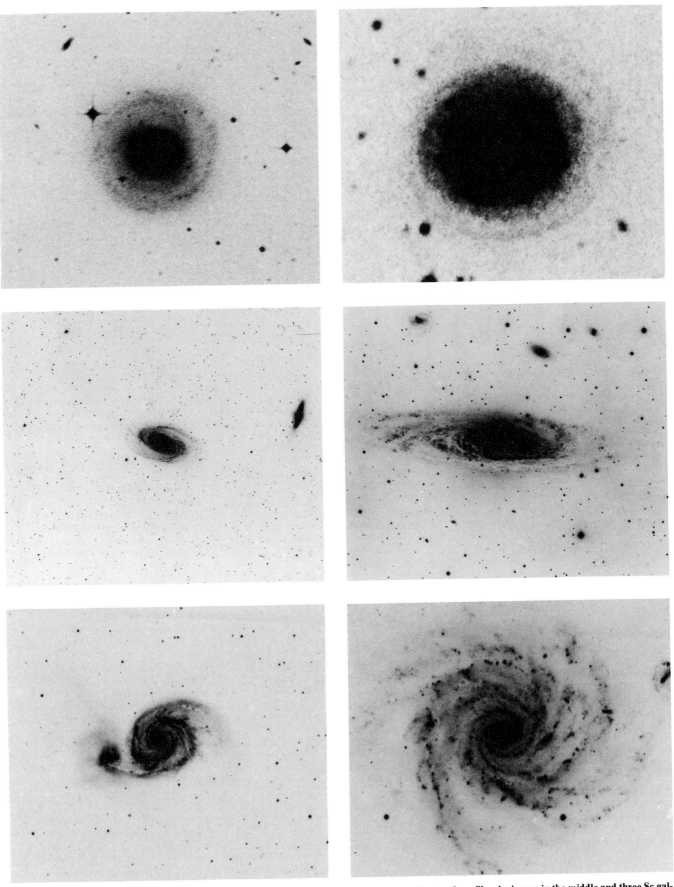

class the galaxies vary enormously in size and luminosity. Here nine examples are reproduced in negative images in which 1.2 centimeters equals 10 kiloparsecs (32,600 light-years). Three Sa galaxies are at the top, three Sb galaxies are in the middle and three Sc galaxies are at the bottom. In the progression from Sa to Sc the nucleus gets smaller with respect to the disk and spiral structure gets more pronounced.

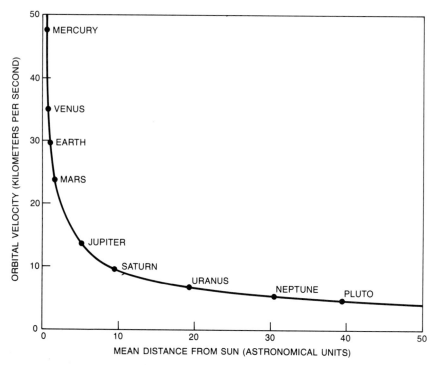

KEPLER'S LAW for the orbital velocity of planets in the solar system, in which more than 99 percent of the total mass resides in the sun, yields this plotted curve. Orbital velocity decreases inversely as the square root of *r*, the planet's mean distance from the sun. The distance is shown here in astronomical units; one A.U. equals the mean distance between the earth and the sun. Pluto, at 39.5 A.U., lies 100 times farther from the sun than Mercury, at .39 A.U. Mercury's orbital velocity is about 47.9 kilometers per second; Pluto's velocity is accordingly slower by a factor of 10, or 4.7 kilometers per second ($47.9 \times 1/\sqrt{100}$). The author's results show that the orbital velocities of stars in a spiral galaxy depart strongly from a Keplerian distribution.

the universe is expanding, carrying almost all the other isolated galaxies away from ours and away from one another. As a result of Smith and Zwicky's work it is known that in pairs, groups and clusters of galaxies the local gravitational field overcomes the general expansion, so that these denser agglomerations of matter remain bound. Although the distances between clusters of galaxies are increasing, the distances between galaxies within clusters remain about the same. Slipher also noted that the spectra of individual galaxies can yield additional information about the motions of stars and gas within the galaxy.

If the disk of a spiral galaxy is oriented so that its plane is sharply tilted with respect to the line of sight from the earth, the rotation of the galaxy will carry the stars and gas on one side of the galactic nucleus toward our galaxy and those on the other side away from it. The spectral lines of the approaching material will therefore be blue-shifted, or raised in frequency, and the lines of the receding material will be red-shifted, or lowered in frequency. A measurement at any point on a spectral line will therefore supply both the angular distance of that point from the galactic nucleus and the velocity along the line of sight at that distance.

It is difficult to make spectroscopic measurements of the velocities of individual stars, which are faint even in galaxies fairly close to our own. In our work, therefore, we observe not stars but the light from the clouds of gas, rich in hydrogen and helium, that surround certain hot stars. The spectra of such clouds consist of bright emission lines that arise as an electron in an excited atom drops from a higher energy state to a lower one. In addition to emission lines of hydrogen and helium there usually are bright lines from atoms of nitrogen and sulfur that are singly ionized, or stripped of one electron. These lines are called forbidden because they arise only from atoms in the near-vacuum of space; in terrestrial laboratories such singly ionized atoms are rapidly deexcited by collisions with other atoms before the forbidden transition can occur.

Until recently it was not possible to get high-resolution optical spectra of the faint outer regions of galaxies. It is the present availability of large optical telescopes, of high-resolution, long-slit spectrographs and of efficient electronic imaging devices that have made our observing program feasible. Six years ago my colleagues and I set out to measure the rotational velocities com-

pletely across the luminous disk of suitably tilted spiral galaxies. Our aim was to study the internal dynamics and distribution of mass in individual galaxies as a function of the galaxies' morphology. We have now observed 60 spiral galaxies: 20 each of the three major types Sa, Sb and Sc. We have selected galaxies that have a well-defined type, that are well inclined to the plane of the sky (yielding a large component of orbital velocity along the line of sight), that have an angular diameter no larger than the slit of the spectrograph and that span a large range of luminosities within each type.

Most of the spectra have been obtained with two four-meter telescopes, the one at the Kitt Peak National Observatory in Arizona and the one at the Cerro Tololo Inter-American Observatory in Chile. A few of the spectra were recorded with the 2.5-meter telescope at the Las Campanas Observatory in Chile.

After the photons from the galactic source pass through the slit of the spectrograph and are dispersed by a diffraction grating, they are focused on a "Carnegie" image tube (RCA C33063), where they are multiplied by a factor of 10 or more before they are recorded by the photographic emulsion. Exposures of two to three hours are recorded on Kodak IIIa-J plates whose sensitivity, matched to that of the image tube, has been much increased by having previously been baked at 65 degrees Celsius for two hours in a special "forming" gas (nitrogen with an admixture of 2 percent hydrogen) and preexposed to flashes of light. Without the image tube and the plate-sensitizing methods exposure times would have been prohibitively long: from 20 to 60 hours.

Generally two exposures are made of each galaxy. In one exposure the spectrograph slit is made to coincide with the major (long) axis of the galaxy; each point on the spectrum arises from a single region of the galactic disk. The Doppler, or velocity, displacements of the emission lines are readily discerned in the developed image. A second exposure is made with the spectrograph slit aligned with the minor axis of the galactic disk. Since the orbital velocities are now perpendicular to the line of sight, no Doppler shifts are evident. The absence of line displacements with the slit along the minor axis is confirming evidence that the motions we study are indeed orbital ones.

In order to have a reference scale against which to measure the displacement of emission lines in galactic spectra astronomers formerly recorded neon lines from a lamp along the edges of the spectrum. We have now dispensed with this procedure. Instead we measure displacements directly from the unshifted lines on each plate that are emitted by hydroxyl (OH) molecules in the earth's atmosphere. Many astronomers have adopted sophisticated plate-scanning devices to measure line positions, particularly for faint signals. We, however, still measure the location of the emission lines with the aid of a microscope whose stage can be moved in two directions. We are able to measure positions in each coordinate to the high accuracy of one micrometer.

In our work we define the nominal radius of a galaxy as that distance at which the surface brightness of the gal-

axy has fallen to the threshold of detectability on plates made with the 48-inch Schmidt telescope on Palomar Mountain, a value equal to 25th magnitude per square second of arc. For establishing the distance to the objects examined, and hence their actual size, we adopt a value for the Hubble constant (which specifies the expansion rate of the universe) of 50 kilometers per second per megaparsec. (A megaparsec is 3.26 million light-years.)

From the measured velocities of the strongest emission lines we compute

a smooth rotation curve by averaging together the approaching and receding velocities from the two sides of the galactic disk. Although each galaxy exhibits distinctive features in its rotational pattern, the systematic trends that emerge are impressive. With increasing luminosity galaxies are bigger, orbital velocities are higher and the velocity gradient across the nuclear bulge is steeper. Moreover, each type of galaxy displays characteristic rotational properties. For example, the most luminous Sa galaxies rotate more than 50 percent faster at the midpoint of their radi-

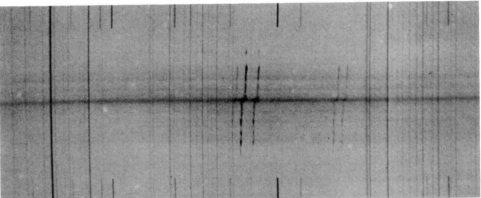

SPECTRUM OF SPIRAL GALAXY NGC 7541 (*right*) was recorded with the four-meter telescope at the Kitt Peak National Observatory by the author and W. Kent Ford, Jr. NGC 7541 is a type Sc spiral, 60 megaparsecs distant. (A megaparsec is 3.26 million light-years.) The exposure time was 114 minutes. The galaxy is seen at the left as it appears on a television monitor in the telescope's console room. The dark line through the galaxy shows the orientation of the spectrograph slit. Light from across the disk is sampled (*see illustration below*).

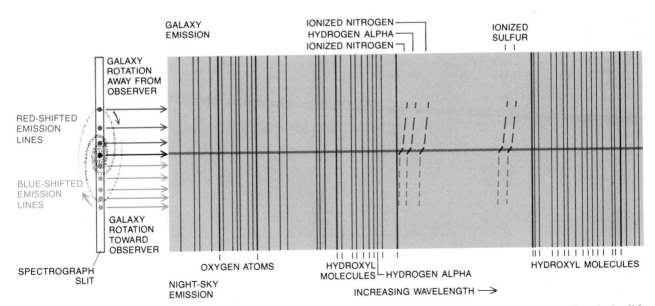

EMISSION LINES in the spectrogram of NGC 7541 arise from two sources: the night sky and atoms in the gas clouds surrounding bright stars in the galaxy. Most of the night-sky lines, which extend across the entire width of the spectrogram, are from hydroxyl (OH) molecules in the atmosphere of the earth. A few arise from oxygen and hydrogen atoms in the earth's atmosphere. The rotation of NGC 7541 shifts the position of the emission lines from the disk of the galaxy to either a shorter (bluer) wavelength or a longer (redder) one, depend-

ing on whether the rotation is carrying the stars and gas in the disk toward or away from the observer. Because the galaxy itself is traveling away from the observer as part of the general expansion of the universe, the hydrogen-alpha line from gas in the galaxy is red-shifted from the position of the same line in the night sky. The displacement is a measure of the galaxy's velocity of recession. The slant of the galactic emission lines shows that the orbital velocity of stars and gas in the disk is increasing with distance from the galactic center.

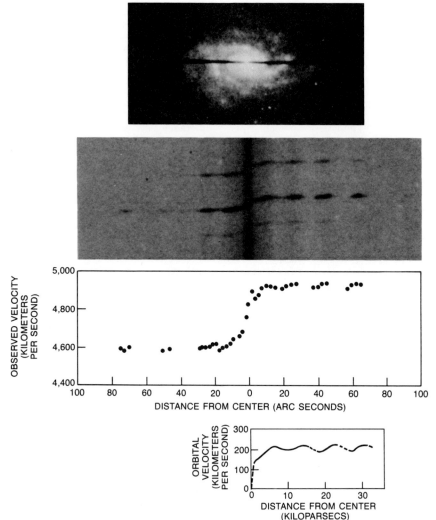

the radius, the density of matter in successive shells must decrease as 1 over the radius squared in order for the product of the density times the volume to remain constant.

The theoretical model that least disturbs generally accepted ideas about galaxies accounts for the observed rotation curves by embedding each spiral galaxy in a spherical "halo" of matter that extends well beyond the visible limits of the galactic disk. The gravitational attraction of this unseen mass keeps the orbital velocities of the galaxies from decreasing with distance from the galactic center. It is perhaps disappointing that the observations yield almost no information on the detailed distribution of the invisible dark matter. One can nonetheless say that the dark matter is not part of the overall background density of matter in the universe but rather is strongly clumped around galaxies. This is evident because the density of nonluminous matter decreases, albeit slowly, with distance from the galactic center, and the density even at large radial distances is between 100 and 1,000 times higher than the mean density of the universe.

Although there are other models that try to account for the high orbital velocities, all are less satisfactory than a single halo of dark matter. If all the required unseen matter is put in a disk, the disk will quickly become unstable and form itself into a bar. The important finding that halos are necessary for stabilizing a disk was first elucidated by Jeremiah P. Ostriker and P. J. E. Peebles of Princeton University.

The observed dynamic effects are reproduced by models of spiral galaxies that put the mass in a nucleus, a surrounding bulge, a disk and a halo. Particularly interesting models have been developed by John N. Bahcall and Raymond M. Soneira of the Institute for Advanced Study, Maarten Schmidt of Cal Tech and S. Casertano of the Scuola Normale Superiore in Pisa. Perhaps the most radical idea for explaining the observed high rotational velocities is one advanced independently by Joel E. Tohline of Louisiana State University and M. Milgrom and J. Bekenstein of the Weizmann Institute of Science. They have proposed that at great distances the Newtonian theory of gravitation must be modified, thereby allowing rotational velocities in galaxies to remain high at such distances from the galactic center even in the absence of unseen mass.

Additional evidence on the high rotational velocities of matter in spiral galaxies is provided by the 21-centimeter radio waves emitted by the neutral (unionized) hydrogen in the galactic disk. Early studies of the 21-centimeter radiation of a few spiral galaxies by Morton

MEASUREMENT OF THE ROTATION of NGC 2998, an Sc galaxy at a distance of 96 megaparsecs in the constellation Ursa Major, begins with the making of a spectrogram. The picture at the top shows the galaxy and superposed spectrograph slit as they appear on a television monitor at the four-meter Kitt Peak telescope. Below it is the hydrogen-alpha region of the spectrogram that resulted from an exposure of 200 minutes. The plotted points depict velocities across the galactic disk as measured from the hydrogen-alpha line. The entire galaxy is receding at 4,800 kilometers per second; the left side of the galaxy is approaching, the right side receding. The last step is to draw a rotation curve by smoothing velocities from both sides of the disk and translating angular distance on the sky into linear distance in the galaxy.

us than equally luminous Sc galaxies. Among Sc galaxies the most luminous rotate more than twice as fast at comparable radial distances as Sc galaxies that are only a hundredth as luminous.

One overwhelming conclusion emerges from our observations. Virtually all the rotation curves are either flat or rising out to the visible limits of the galaxy. There are no extensive regions where the velocities fall off with distance from the center, as would be predicted if mass were centrally concentrated. The conclusion is inescapable: mass, unlike luminosity, is not concentrated near the center of spiral galaxies. Thus the light distribution in a galaxy is not at all a guide to mass distribution.

On the basis of their rotational velocities the masses of the galaxies in our study range from 6×10^9 to 2×10^{12} times the mass of the sun inside their optical radius. We cannot yet specify the total mass of any one galaxy because we do not see any "edge" to the mass. Instead the mass inside any given radial distance is increasing linearly with distance and, contrary to what one might expect, is not converging to a limiting mass at the edge of the visible disk. The linear increase of mass with radius indicates that each successive shell of matter in the galaxy must contain just as much mass as every other shell of the same thickness. Since the volume of each successive shell increases as the square of

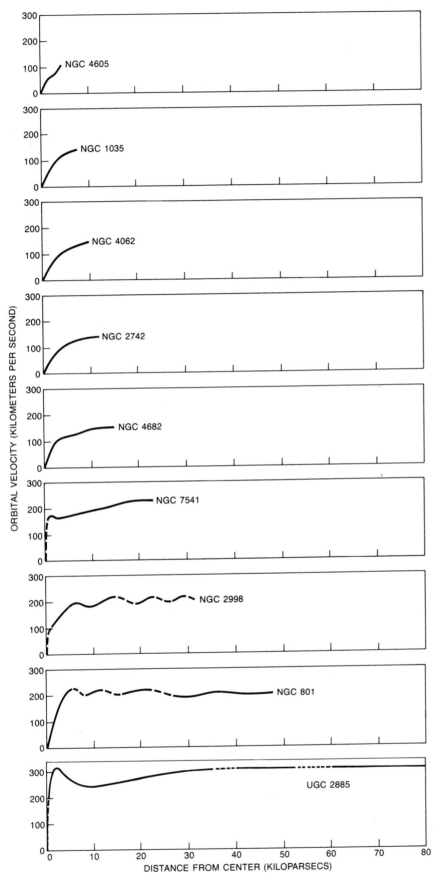

ROTATION CURVES show orbital velocities of nine Sc galaxies from the center outward. Galaxies increase in luminosity from top to bottom. With increasing luminosity galaxies are larger, orbital velocities are higher and velocity gradients near the galactic center are steeper.

S. Roberts of the National Radio Astronomy Observatory showed that the rotational velocities of the hydrogen are high. With multiple radio telescopes, notably the array at Westerbork in the Netherlands and the Very Large Array at Socorro, N.M., it is possible to match and even exceed the resolving power of optical telescopes and thereby to study the distribution of hydrogen in galaxies similar to those we have observed. Albert Bosma of the State University of Leiden has shown for a wide variety of galaxy types that the orbital velocities of neutral hydrogen remain high at large distances from the galactic center.

In general the apparent diameters of galaxies are similar whether they are measured by optical observations or by radio ones. For a small set of galaxies, however, hydrogen extends several times as far out from the center as the luminous stars do. For such objects it is possible to determine the gravitational potential beyond the limits of the optically visible galaxy. In several instances the hydrogen does not remain in a plane but is warped sharply near the edge of the visible disk. It is therefore not certain whether the gas velocities that have been measured at the largest distances from the center are true circular orbital velocities or whether they represent more complex motions.

Renzo Sancisi of the University of Groningen, who has studied such warped galaxies, has suggested that the orbital velocities may in fact be decreasing beyond the limits of the visible galaxy. The velocities, however, seem to decrease only slightly, perhaps by 20 kilometers per second, or about 10 percent, and then hold constant at that value at larger distances. The radio observations are continuing and should offer important information on the far outer regions of galaxies.

Students of galaxies are fortunate in being able to examine the properties of galaxies a long way off and then return to the galaxy where they live and ask if it exhibits the same features as other galaxies. It was not so long ago that astronomers believed the sun, about eight kiloparsecs from the center of our galaxy, was near the edge of it and that the galaxy was only of moderate size. Now all the evidence indicates that our galaxy too extends well beyond the position of the sun and that its mass continues to increase.

The velocity of the sun in its orbit around the center of the galaxy is placed at 220 kilometers per second by James E. Gunn and Gillian R. Knapp of Princeton and Scott D. Tremaine of the Massachusetts Institute of Technology. Other estimates run as high as 260 kilometers per second. At the lower value the amount of mass between the sun and

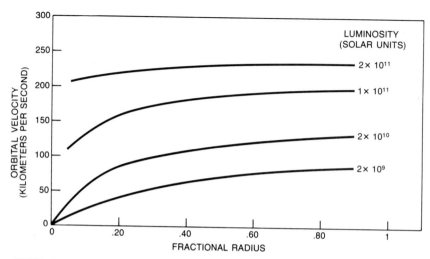

ORBITAL VELOCITIES are depicted schematically for Sc galaxies of varying luminosity as a function of the optically visible radius of each galaxy. Luminosities in solar units differ by two orders of magnitude. At every radial distance orbital velocities increase with luminosity.

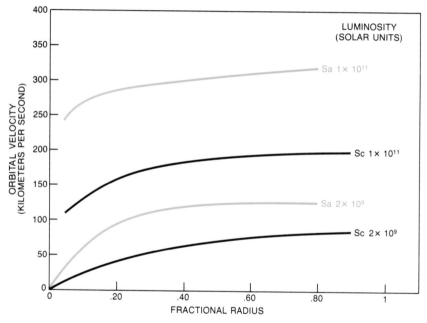

COMPARISON OF Sa AND Sc GALAXIES shows that for equal luminosity orbital velocities are significantly higher in Sa galaxies than they are in Sc galaxies at every radial distance. This implies that Sa galaxies harbor more mass per unit of luminosity than Sc galaxies do.

the center of the galaxy is about 10^{11} solar masses. On the evidence that substantial mass lies beyond the sun's distance from the galactic center, the galactic mass out to 100 kiloparsecs may reach 10^{12} solar masses, which would place our galaxy in a class with the largest galaxies of its type.

Some 30 years ago Jan H. Oort of the Leiden Observatory demonstrated that the observable mass of stars and gas in the galactic disk in the vicinity of the sun is too low by almost a factor of two to account for the disk's gravitational attraction on the stars far out of its central plane. This study offered the first evidence that our galaxy too harbors mass that is not luminous.

More recent evidence comes from the orbital velocities of objects in the plane of the galaxy considerably farther out than the sun. Measurements are difficult, but the velocities have been deduced for a few special cases. For example, Leo Blitz of the University of Maryland at College Park has determined the velocities of clouds of carbon monoxide at a distance of nearly 16 kiloparsecs from the galactic center. These velocities, together with the velocities of hydrogen clouds determined by Blitz and Shrinivas Kulkarni and Carl E. Heiles of

the University of California at Berkeley, yield a rotation curve that continues to rise with increasing distance from the galactic center.

In order to deduce the mass at still larger distances the velocities of globular star clusters in the halo of our galaxy, with one sample of clusters at 30 kiloparsecs from the center and another at 60 kiloparsecs, have been measured by F. D. A. Hartwick of the University of Victoria, Wallace L. W. Sargent of Cal Tech, Carlos Frenk of the University of Cambridge and Simon White of the University of California at Berkeley. Their work shows that the mass continues to increase with approximate linearity to the mean distance of the clusters.

With effort and imagination it is possible to sample the gravitational potential at even more remote distances. Our galaxy is not alone in intergalactic space; it has a retinue of smaller satellite galaxies. The orbits of the two closest satellites, the Large and Small Clouds of Magellan, a little less than 60 kiloparsecs from the center of our galaxy, are highly uncertain. Model orbits have been proposed, however, by Tadayuki Murai and Mitsuaki Fujimoto of Nagoya University, D. N. C. Lin of the Lick Observatory and Donald Lynden-Bell of the University of Cambridge. From the model orbits they deduce values of mass that are consistent with those yielded by the globular clusters.

For still greater distances Jaan Einasto and his colleagues at the Estonian S.S.R. Academy of Sciences have relied on a combination of enormously distant globular clusters and satellite galaxies to deduce the mass to distances beyond 80 kiloparsecs. When the results from such analyses are combined, they indicate a galaxy in which orbital velocities remain in the range of 220 to 250 kilometers per second out to almost 10 times the distance of the sun from the galactic center. Such a mass distribution is mandatory if our galaxy is to resemble all the other spiral galaxies my colleagues and I have studied. It moves the sun from a relatively rural position to a much more urban one.

The broad conclusion that can be drawn from all these results is that as the disk of a spiral galaxy is scanned from the center outward the total mass of luminous and dark material falls off slowly and the luminosity (measured in the blue region of the spectrum) falls off rapidly. As a result the ratio of the local mass density to the local (blue) luminosity density, which can be expressed for convenience as the value of the ratio M/L, increases steadily with distance from the galactic center. In the nuclear region a lot of luminosity is produced by relatively little mass, whereas at large distances little luminosity is produced by a lot of mass. If there were no invisible material clumped around galaxies, the mass distribution would simply follow the luminosity distribution and the M/L ratio would be approximately constant across the disk from its center to its edge.

If mass and luminosity are measured in units of solar mass and solar luminosity, the M/L ratio of the sun is of course 1/1. In such units (omitting the denominator, which is simply 1) the average M/L ratio near the nucleus of a spiral galaxy has sunlike values of 1 or perhaps 2 or 3. Toward the edge of the visible disk, as luminosity decreases, the M/L value climbs to 10 or 20. Beyond the visible disk, where the luminosity falls essentially to zero and the mass remains high, the average M/L value soars into the hundreds.

In an effort to identify the constituents of the invisible halo we must ask what celestial objects have high values of M/L. Stars like the sun are clearly ruled out. The luminous hot young stars that delineate a galaxy's spiral arms are even poorer candidates; their M/L values are about 10^{-4}. At the other extreme the old red-dwarf stars that populate the nuclear bulge and the outlying regions of the galaxy have both a low mass and a low blue luminosity. Their M/L values, about 20, are still far short of the values needed for the halo. Moreover, a halo consisting of very-low-mass red dwarfs would reveal its presence by radiating strongly in the infrared region of the spectrum. All attempts to detect a halo by its visual, infrared, radio or X-ray radiation have failed.

What candidates are left? Normal stars radiate energy generated by thermonuclear processes, which convert hydrogen and helium into heavier elements. Such nuclear processes are kindled only in bodies whose mass is large enough for the gravitational energy to raise the temperature at the core of the star to several million degrees Kelvin (degrees C. above absolute zero). The minimum mass required is about .085 times the mass of the sun. Jupiter, the largest planet in the solar system, falls short of this value by a factor of nearly 100. A halo of planetlike bodies, perhaps protostars that failed to become stars, is at least conceivable, although rather unlikely. In sum, the only requirement for the halo is the presence of matter in any cold, dark form that meets the M/L constraint, from neutrinos to black holes.

So far I have described the rotational properties of relatively isolated normal spiral galaxies. There is additional observational evidence for high M/L ratios at large distances from the nuclei of

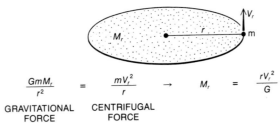

$$\frac{GmM_r}{r^2} = \frac{mV_r^2}{r} \rightarrow M_r = \frac{rV_r^2}{G}$$

GRAVITATIONAL   CENTRIFUGAL
FORCE   FORCE

	NGC 1035		NGC 2998	
RADIUS (KILOPARSECS)	VELOCITY V_r (KILOMETERS PER SECOND)	INTERIOR MASS M_r (10^{10} SOLAR MASSES)	VELOCITY V_r (KILOMETERS PER SECOND)	INTERIOR MASS M_r (10^{10} SOLAR MASSES)
.5	39	.018	87	.088
1	65	.098	102	.24
2	91	.39	126	.74
3	107	.80	142	1.4
5	123	1.8	182	3.9
8	135	3.4	204	7.7
20			214	21
30			214	32

MASS INSIDE A GIVEN RADIAL DISTANCE can be calculated from the equivalence of gravitational force and centrifugal force at distance r from the center of the galaxy. In the equations G is the constant of gravitation, m is the mass at distance r, M_r is the mass inside r and V_r is the orbital velocity of mass m. The mass inside r increases linearly with distance. The table gives the mass inside r for two Sc galaxies: NGC 1035, of low luminosity, and NGC 2998, of high luminosity. At every distance from the galactic center the more luminous galaxy exhibits a higher orbital velocity and therefore must have much more mass inside that distance.

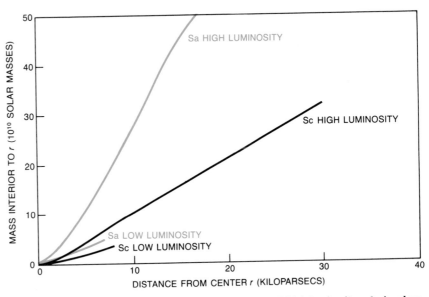

COMPARISONS OF INTERIOR MASS for both low- and high-luminosity galaxies show that the mass rises with approximate linearity with distance r from the center and gives no sign of approaching a limit at the edge of the optically visible galaxy. At every radial distance Sa galaxies exhibit higher mass and therefore higher density than Sc galaxies of equal luminosity.

other galaxies. Occasionally nature offers an unexpected opportunity to probe its secrets. Recently François Schweizer of the Carnegie Institution, Bradley C. Whitmore of Arizona State University and I have been fascinated by the faint "anonymous" galaxy AO 136–0801, one of a class of spindle galaxies

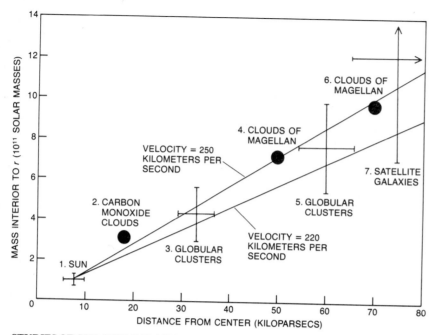

STUDIES OF OUR OWN GALAXY yield estimates of the mass inside *r* measured in kiloparsecs from the galactic center. The estimates are calculated from the orbital velocities and mean distances of a variety of objects. The value adopted for the orbital velocity of the sun at a distance of eight kiloparsecs is 220 kilometers per second. The second point is obtained from the mean velocity of carbon monoxide clouds at a mean distance of 18 kiloparsecs, measured by Leo Blitz of the University of Maryland at College Park. The third and fifth points are derived from the velocity of globular clusters of stars in the "halo" of our galaxy at two different average distances from the nucleus. The velocities of the nearer clusters were analyzed by Carlos Frenk, who was then working at the University of Cambridge, and Simon White of the University of California at Berkeley. The velocities of the more distant clusters were studied by F. D. A. Hartwick of the University of Victoria and Wallace L. W. Sargent of the California Institute of Technology. The fourth point was obtained from velocities of the Clouds of Magellan, the nearest galaxies to our own, as estimated by Tadayuki Murai and Mitsuaki Fujimoto of Nagoya University. The sixth point represents independent estimates of the distance and velocities of the Clouds of Magellan made by D. N. C. Lin of the Lick Observatory and Donald Lynden-Bell of the University of Cambridge. The final point is based on the velocity of more remote satellite galaxies as estimated by Jaan Einasto and his colleagues at the Estonian S.S.R. Academy of Sciences. The extent of the vertical lines indicates the range of values for orbits of different geometries. The measurements suggest that rotational velocities in our galaxy lie between 220 and 250 kilometers per second and remain constant out to 80 kiloparsecs, or roughly 10 times the sun's distance from the galactic center. The mass inside 80 kiloparsecs is likewise some 10 times the mass inside the radial distance of the sun, or about 10^{12} solar masses.

right angles to the plane of rotation of the disk. It seems improbable that this dynamical configuration could have arisen in the normal evolution of an isolated disk galaxy; the configuration must be the result of some kind of event, such as an encounter with another galaxy or with a disk of gas. By measuring the displacement of emission lines we find that the ring's velocity of rotation is about 170 kilometers per second and that the velocity curve is flat or slightly rising out to a distance of almost three times the radius of the inner disk. If the velocity curves of the disk and the ring are plotted on the same velocity-distance scale, the two are seen to have nearly identical values at the same distance from the center of the galaxy. The high rotational velocity of the ring offers strong evidence for the existence of a massive halo extending at least three times farther than the visible radius of the disk. Moreover, the shape of the halo must be more nearly spherical than disklike. Calculations show that if the halo were as flat as the disk, the velocities above the plane of the disk would be smaller than those in the disk by 20 to 40 percent.

I have been describing determinations of mass made by measuring the velocity of orbiting test objects, objects in the central disk of a galaxy and objects orbiting the pole of an unusual galaxy. Other special instances can help to shed light on the quantity of dark matter in the universe. Galaxies often exist in pairs. In such instances one galaxy can be considered a test object in orbit around the other. The analysis of such a system is complex because both the orientation of the orbit in space and the position of the galaxy in the orbit are unknown. One can, however, resort to the observed properties in a large sample of double galaxies (the difference between the velocities of the two galaxies, their angular separation and their luminosity) to infer from statistical arguments the probable distribution of orbital elements and *M/L* ratios appropriate to the galaxies.

Independent analyses by Edwin L. Turner of Princeton, Steven D. Peterson, working at Cornell University, Linda Y. Schweizer of the Carnegie Institution and I. D. Karachentsev of the Special Astrophysical Observatory in the U.S.S.R. yield mean *M/L* values in the range between 25 and 100. These values of *M/L* are an average over a distance equal to the separation of the galaxies in each pair, a distance generally equal to several galaxy diameters, or on the order of 100 kiloparsecs. This result helps to confirm the view that halos of dark matter, with large values of *M/L*, extend well beyond the optical limits of galaxies.

We can now return to our original

with polar rings. It is called anonymous because it is not listed in any of the standard galactic catalogues; its numerical designation corresponds to its location in the sky.

Our observations of the distribution of light across the spindle show that it is a low-luminosity disk of stars viewed nearly edge on, with little or no gas and dust and no spiral structure. Such galaxies are classed as SO galaxies and represent a significant fraction of all disk galaxies. By our usual methods we have determined the rotational properties of the disk by measuring the Doppler shift of absorption lines from its component stars. A short distance from the center of the object along the major axis of the spindle rotational velocities reach 145 kilometers per second, a value that corresponds closely to velocities measured in type-Sa galaxies of low luminosity. Along the minor axis the orbital velocities show no line-of-sight component,

confirming evidence that we are observing a rotating disk of stars.

The unusual feature of AO 136–0801 is a large ring, also seen nearly edge on, that encircles the narrow axis of the spindle by passing almost over the disk's center of rotation [*see illustration below*]. The ring is composed of gas, dust and luminous young stars. The gas reveals itself by its emission-line spectrum, the dust by its absorbing effects where it crosses in front of the spindle and the stellar component by its knotty, bluish appearance in photographs. The maximum diameter of the ring is several times greater than the long axis of the spindle. As a consequence the motions of the objects in the ring offer a unique opportunity to probe the gravitational field perpendicular to the galactic disk out to distances exceeding the visible radius of the disk.

Our spectrographic observations confirm that the ring is indeed rotating at

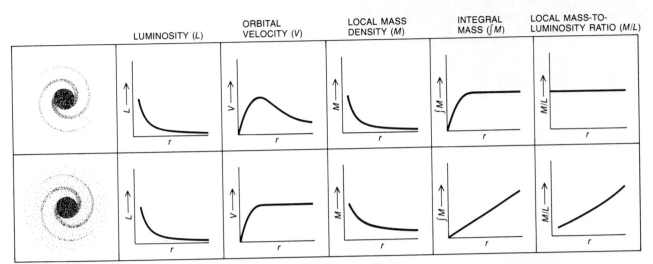

| LUMINOSITY (L) | ORBITAL VELOCITY (V) | LOCAL MASS DENSITY (M) | INTEGRAL MASS (∫M) | LOCAL MASS-TO-LUMINOSITY RATIO (M/L) |

HYPOTHETICAL AND ACTUAL GALAXIES deviate sharply in all their properties except luminosity. The typical actual spiral galaxy at the bottom has a massive nonluminous halo. The hypothetical galaxy at the top has no halo. Its surface brightness decreases rapidly, orbital velocities outside the nucleus decrease in Keplerian fashion, local mass density falls in parallel with luminosity, integral mass reaches a limiting value and the ratio of mass to luminosity stays approxi- mately constant with increasing radial distance. Such were the expect- ed properties of a galaxy. In an actual galaxy the presence of a dark halo changes everything but the galaxy's optical appearance. The or- bital velocities remain high, the local mass density falls only slowly, the integral mass increases linearly with radius and the mass-to-lumi- nosity ratio steadily increases as the halo of the galaxy contributes more mass and the luminous disk falls to the threshold of detectability.

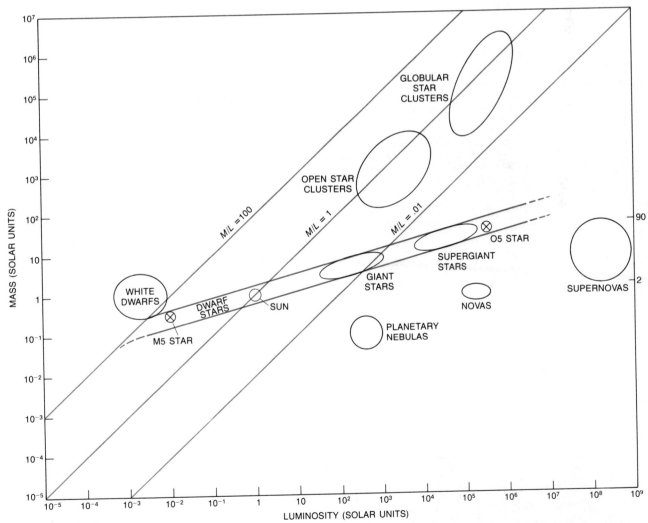

MASS AND LUMINOSITY ARE COMPARED for typical com- ponents of a spiral galaxy such as our own. The mass and luminosity of the sun are taken as unity. In solar units the value of the ratio of mass to luminosity, M/L, for normal stars decreases from about 30 for cool, old dwarf stars (type M5) to about 10^{-4} for hot, young stars (type O5). Only extremely dense white-dwarf stars have an M/L val- ue in excess of 100. Some other class of objects is needed to popu- late the halo of a galaxy, where M/L values soar into the hundreds.

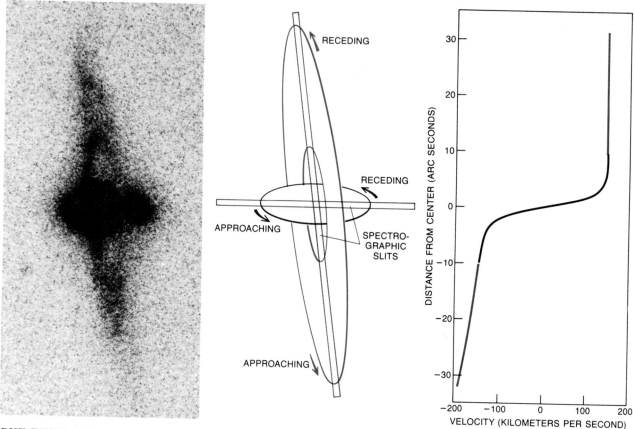

DISK INSIDE A RING is exhibited by the "anonymous" galaxy AO 136–0801, so called because it is not listed in standard catalogues. (The numbers give its position in right ascension and declination.) The oval central region is a rotating disk of stars seen nearly edge on. The stars and gas in the thin ring are also rotating but in a plane almost perpendicular to the disk. The directions of rotation are indicated by the diagram in the middle, which shows how the slits of the spectrograph were oriented for measuring orbital velocities in the disk and the ring. The two sets of velocity measurements are plotted at the right. At 10 seconds of arc from the center the velocities in the disk (*black*) and in the ring (*color*) are essentially the same. The velocities in the ring, however, can be measured out to nearly three times the optical radius of the disk, and they remain virtually constant. It appears that the mass is continuing to increase linearly out to distances much greater than the disk radius, and that objects in the ring respond to a gravitational potential that is not disklike but spherical.

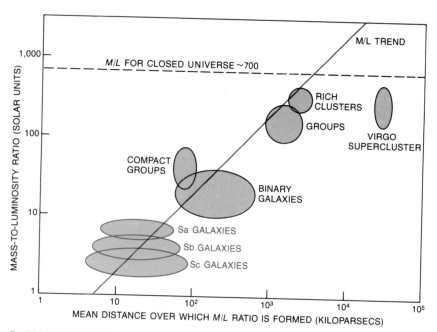

RATIOS OF MASS TO LUMINOSITY are plotted for aggregates of matter on progressively larger scales. The plot is based on one devised by Herbert J. Rood of the Institute for Advanced Study. The *M/L* value of a density of matter sufficient to arrest the expansion of the universe is about 700. For galaxies values are below 10. The value rises with the size of the aggregate.

question: Does the universe contain enough invisible matter to raise the average density to 5×10^{-30} gram per cubic centimeter, the value needed to close the universe and bring its expansion to a halt? As we have seen, such a density would be reached if the density of nonluminous matter exceeded the density of luminous matter by a factor of about 70. Alternatively what would be needed to close the universe can be expressed in terms of the ratio of total mass to luminosity. That value is roughly 700, compared with 1 for the sun.

Is there any evidence that the *M/L* value of 700 is approached? Averaged over the visible disks of spiral galaxies, the ratio of total mass (luminous and nonluminous) to luminosity is about 5. For SO and elliptical galaxies the *M/L* value is higher, on the order of 10. For double galaxies and small groups of galaxies the *M/L* value increases to between 50 and 100. Analyses of galaxy motions in large clusters indicate *M/L* values of several hundreds. This increase in mean *M/L* value with increasing distance from the center of the system was first stressed a decade ago by Einasto, Ants Kaasik and Enn Saar of the Estonian S.S.R. Academy of Sciences, and also by Ostriker and Peebles and by Amos Yahil of the State University of New York at Stony Brook. So far there is no evidence for the existence of M/L values above the critical one of 700 needed to close the universe. The highest of the derived values, however, comes tantalizingly close. Some physicists consider it significant that the inferred values seem to be converging on the critical one rather than being orders of magnitude either higher or lower.

Investigations encompassing gigantic distances and vast time scales are made more difficult by this new realization that the distribution of light is an unreliable guide to the distribution of mass in the universe. An unknown fraction of the mass in a spiral galaxy is hidden in a nonluminous constituent, and so is an unknown fraction of the mass in clusters of galaxies. One cannot yet state whether regions of the universe that are devoid of galaxies are simply voids of light or are voids of mass as well. To answer this question astronomers will have to be clever in devising novel observing techniques and physicists will have to determine the properties of exotic forms of matter. Only then will it be possible to establish the nature of the ubiquitous dark matter, to determine the full dimensions and mass of galaxies and to assay the likely fate of the universe.

The Evolution of Disk Galaxies

by Stephen E. Strom and Karen M. Strom
April, 1979

A spiral galaxy can evolve into a smooth disk without spiral arms. Whether or not it does so depends on its environment: the galaxy most likely to evolve into a smooth disk belongs to a rich cluster

Ever since galaxies were first recognized as being "island universes" outside our own Milky Way galaxy they have been utilized as probes of the large-scale structure and evolutionary history of the universe. Their usefulness for this purpose depends critically on knowledge of when the galaxies were formed and how their bulk properties (size, luminosity and color) and detailed structure have changed with time. The development of plausible models of galaxy formation and evolution has therefore presented astronomers with a stimulating challenge. Developing such models is particularly difficult because most galaxies were evidently born some 15 billion years ago and evolved most rapidly at a time well before the present epoch. Hence there are no "snapshots" of nearby galaxies in a variety of evolutionary stages.

How then can one develop the observational basis for shaping and testing models of galaxy formation and evolution? First, one can test the models against the observable characteristics of the nearby, highly evolved galaxies that are most amenable to detailed study. Second, one can look back in time by observing galaxies located at distances of several billion light-years and comparing their properties with those characteristic of nearby systems. Third, one can search for nearby galaxies that are young or that appear to have evolved slower than the typical systems.

Recent studies of disk galaxies illustrate the usefulness of all these approaches. A disk galaxy has two morphologically distinct parts: a central bulge consisting of a spheroidal aggregation of stars and a surrounding disk consisting of stars fanning outward in a thin layer. The relative sizes of the bulge and the disk vary from a nearly pure bulge in some galaxies to a nearly pure disk in others. The bulge region of most disk galaxies seems to be totally lacking in young stars. The disk region, however, varies considerably in star-forming activity. In disk galaxies of the spiral type newly formed stars and their associated complexes of ionized hydrogen gas define the spiral arms that show up spectacularly in photographs. In disk galaxies of the SO type, on the other hand, the disks are smooth and devoid of young stellar complexes. Furthermore, the disks of SO galaxies show no evidence of the gas required to fuel star formation in the future.

Spiral galaxies seem to inhabit environments that are significantly different from those of SO galaxies. Whereas spirals are the dominant type of system in regions where galaxies are sparse and widely spaced, they are rare in the central regions of the densely populated great clusters of galaxies, which may consist of from a few hundred to more than 1,000 individual galactic systems in a volume of space ranging from a million to 10 million light-years across. SO galaxies, in contrast, are by far the commonest type of disk system in the great clusters and are less abundant than spirals in regions of lower galactic density.

Current studies of galactic evolution are aimed at isolating the genetic and environmental factors that control the appearance of galaxies as a function of time. What accounts for the relative prominence of the bulge and disk components? Why do stars continue to form in spiral galaxies but not in SO galaxies? What accounts for the difference in the frequency distribution of spirals and SO's in regions with different densities of galaxies? We shall present evidence suggesting that the answers to these questions can be found primarily in the environmental conditions that prevail during the evolution of a galaxy.

Disk galaxies probably begin their life early in the history of the expanding universe as roughly spherical, rotating protogalactic clouds consisting chiefly of clumpy hydrogen and helium gas. When the self-gravity of the cloud overcomes the competing effect of the expansion of the universe, the cloud begins to collapse. The collapse proceeds most rapidly in the central region of the cloud and slower in the outer regions.

When the central region reaches a certain density, stars begin to form, consuming much of the available gas. The result is a spheroidal system between 10,000 and 100,000 light-years in diameter containing between 10^{10} and 10^{12} stars, a system that is outwardly similar to the galaxies designated elliptical.

The low-density gas remaining in the slowly collapsing outer regions of the cloud is so diffuse that in those regions rapid star formation is precluded. Much of the gas therefore remains in unprocessed clumps. When the gas clumps collide, they heat up; their kinetic energy is thereby converted into radiation that escapes from the galaxy. The clumps collide primarily along the axis of rotation of the protogalactic cloud. Eventually the energy in the clump motions parallel to the rotation axis is dissipated in collisions, with the result that the gas finally settles into a rotating disk. When the density of the gas reaches a critical value, stars begin to form. The relative sizes of the disk and the bulge depend on the efficiency of star formation early in the collapse of the protogalactic cloud. If a large fraction of the cloud is initially made into stars, the amount of gas remaining to form a disk will be small. The system will therefore have a large bulge and a small disk.

Instabilities in the newly formed stellar disk lead to the generation of a spiral-wave pattern in the disk stars. A time-sequence recording of the density distribution of stars on the surface of the disk would show density-wave crests appearing to move through the disk at an angular speed called the pattern speed and designated Ω_p (capital omega with a subscript p, for pattern). The disk gas at a given distance, r, from the center of the galaxy moves in orbit around the center at an angular speed, $\Omega(r)$, that typically exceeds Ω_p. Frank H.-S. Shu and William W. Roberts, Jr., of the University of Virginia and their collaborators showed that as the gas flows through the density-wave crests a shock wave is generated if both the quantity $\Omega(r) - \Omega_p$ and the wave amplitude are sufficiently

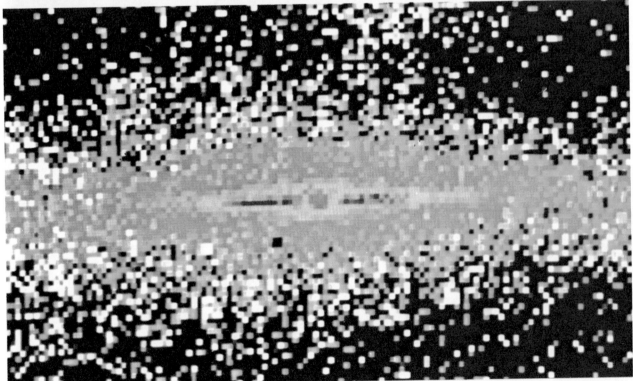

SPIRAL GALAXY NGC 4762, the flattest edge-on galaxy known, is depicted here in two color-coded images created by the authors with the aid of the interactive picture-processing system (IPPS) recently developed at the Kitt Peak National Observatory. For comparison a conventional photograph of NGC 4762, which is 60 million light-years away, appears at the top right in the set of nine galaxies illustrated on the next page. Under computer control, the IPPS presents a digitized picture on a display screen. By typing in simple instructions the astronomer can quickly see the effect of altering brightness levels or choosing color-coding schemes to emphasize features of interest. In the top picture the original photograph of NGC 4762 has been manipulated so that the bluest areas correspond to the faintest regions of the galaxy. The bottom picture is synthesized from two photographs of the same galaxy, one made in the ultraviolet, the other in the red. Galactic regions with the highest ultraviolet-to-red ratios are coded in blue. They can be seen to correspond with the faint (blue-coded) areas in the top picture. The authors believe the blue regions are dominated by old stars that contain only small amounts of elements heavier than helium. In contrast, the stars in the red-coded regions are believed to be rich in heavier elements.

large. The conditions in the compressed region behind the shock wave are believed to be conducive to forcing the collapse and fragmentation of the gas, thereby producing clusters of stars. The luminous, hot but short-lived stars in the newborn clusters define the bright arms that are the most prominent features in photographs of spiral galaxies.

The rate at which star-forming events are induced by the shock waves depends on the frequency with which the gas encounters the crests of the density waves. The frequency is approximately equal to 1 divided by $\Omega - \Omega_p$ and usually is highest toward the inner region of the galaxy and decreases outward. Hence the rate of star formation and the resulting depletion of gas are highest in the inner regions. As the galaxy evolves, gas is exhausted by the formation of stars, first in the central regions and then over an increasingly large fraction of the disk. The most rapid exhaustion of gas is expected in systems with a prominent bulge, a high rate of rotation of the bulge with respect to the disk and a large value of $\Omega - \Omega_p$.

Fuel for the formation of more stars can be provided only by the slow ejection of gas by dying stars in the disk, the infall of gas left over from the formation of the galaxy or the accretion of gas from outside the galaxy. Sandra M. Faber of the University of California at Santa Cruz and John S. Gallagher of the University of Illinois have suggested, however, that such sources of replenishment may be nullified by the effects of a wind of high-velocity gas emanating from the spheroidal bulge region. The energy of the wind would derive from two sources: from the heat generated by supernova explosions and that generated by collisions between the shells of gas ejected by dying stars. If the heating is strong enough to overcome radiative cooling (the energy escaping as photons,

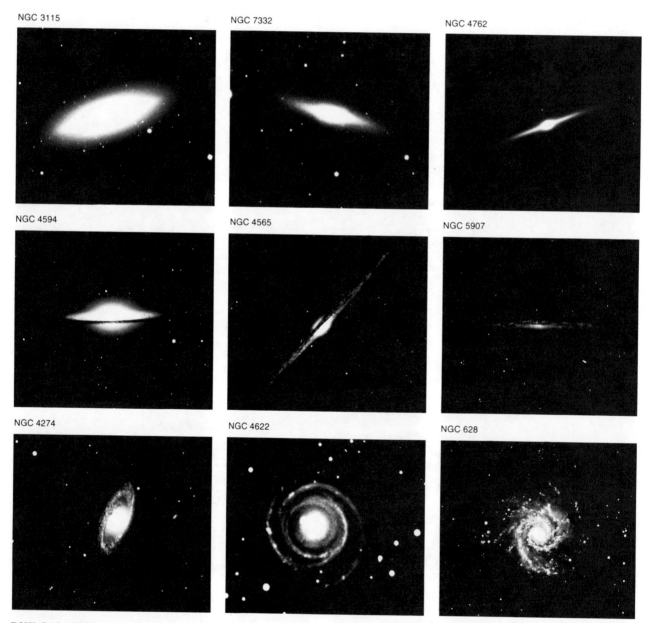

DISK GALAXIES can be divided into two broad categories: spiral systems and SO systems. Both kinds of galaxy have a central bulge and a surrounding disk. The disks of spiral galaxies have visually prominent arms because they are studded with complexes of bright, newly formed stars. The disks of SO galaxies, in contrast, are smooth, show no spiral structure and are devoid of young stars. The three photographs in the top row depict SO systems viewed nearly edge on, arranged in order of the decreasing prominence of their bulge with respect to their disk. No evidence of recent star formation is visible. The photographs in the middle row show three galaxies of the spiral type, also seen edge on and in order of decreasing bulge-to-disk ratio. The galaxies in the bottom row illustrate the probable face-on appearance of spiral galaxies in the middle row. The bright knots in the spiral arms of the galaxies represent newly formed stellar complexes.

or quanta of electromagnetic radiation), the equilibrium temperature of the gas in the bulge will be so high that the gas will no longer be gravitationally bound to the bulge. In general massive bulges will produce the strongest winds. If the pressure of the outflowing wind is high enough, it will deplete the surrounding disk of its gas.

When all the gas has been removed by the formation of stars or the action of galactic winds, the bright spiral arms defined by newly formed stars disappear. The density-wave pattern in the older disk stars, however, remains. The subsequent fate of the density wave in a system free of gas is not well understood. Our best guess is that the wave will at first increase in amplitude and then will damp out within a few galactic rotations, say within half a billion to a billion years. If our assumption is correct, gas-free disk systems will then resemble SO galaxies. The time needed for a spiral galaxy to be transformed into an SO galaxy depends on the amount of gas left after the initial formation of disk stars, the rate of star formation and the rate at which gas is added to or removed from the disk.

Although this picture seems quite plausible, so far there is no convincing observational evidence to support the hypothesis that spiral galaxies evolve naturally into SO systems when all sources of gas have been removed by star formation or galactic winds. The search for such evidence currently centers on attempts to compare the frequency of SO galaxies in a sample of disk systems where the gas-depletion rate should be high with the frequency of such galaxies in a second sample where the gas-depletion rate should be low. If SO galaxies evolve from spirals, one would expect to find the highest frequency of SO galaxies in systems characterized by high gas-depletion rates. If star formation is the dominant mechanism of gas removal in spirals, $\Omega - \Omega_p$ will measure the star-formation frequency and implicitly the gas-depletion rate. Hence SO's should be the dominant type among disk systems that have a large value of $\Omega - \Omega_p$, namely those with a prominent bulge. Spirals should be observed most frequently among galaxies with a lower bulge-to-disk ratio and a smaller value of $\Omega - \Omega_p$.

If galactic winds rather than star formation dominate the depletion of gas at late stages of the evolution of a disk, one would expect the systems with the strongest winds to evolve most rapidly from spiral to SO. Since galaxies with a large bulge-to-disk ratio should have the strongest winds, one would again expect the highest frequency of SO galaxies among disk systems with large bulges. Several investigators are attempting to measure the relative abundance of SO

GIANT ELLIPTICAL GALAXY NGC 4472, representative of systems that have no disk, is presumed to be spheroidal in shape. Like SO galaxies, it is entirely lacking in young stars. Such giant galaxies have a mass about 10^{12} times the mass of the sun, which makes them five or 10 times as massive as our own large spiral galaxy. NGC 4472 is 60 million light-years away.

galaxies in samples of disk galaxies differing in their bulge-to-disk ratio, but so far no definitive results have been presented.

If the removal of gas from a spiral galaxy results in an evolutionary transition to an SO galaxy, what accounts for the greater frequency of such transitions in rich clusters of galaxies? Recent X-ray observations of galaxy clusters may have provided the essential clue to the solution of the mystery. X-ray surveys of the sky in the range of photon energies between 1,000 and 10,000 electron volts by the *Uhuru* satellite reveal a number of luminous X-ray sources associated with rich clusters of galaxies. The X-ray flux appears to be highest in clusters that have the highest total mass, the highest density of galaxies and the highest concentration of galaxies toward the center of the cluster. The X rays are thought to represent the radiation emitted by energetic electrons that are accelerated in a field of positive ions. The emission arises within an intracluster medium that varies in density from about one electron to 10 electrons per 100,000 cubic centimeters elevated to a temperature of about 10^8 degrees Kelvin. The density of the gas is highest in the center of the cluster and decreases

smoothly toward the outer regions of the cluster.

J. Richard Gott III of Princeton University and James E. Gunn of the California Institute of Technology suggest that spiral galaxies moving through the intracluster medium will be subject to a "ram" pressure high enough to remove the disk gas. The ram pressure depends on the density of the gas within the intracluster medium and the velocity of the galaxy with respect to the medium; the pressure is highest for dense mediums and high velocities. From density estimates based on X-ray observations of the rich cluster of galaxies in the constellation Coma Berenices and from velocity measurements of individual cluster galaxies Gott and Gunn conclude that spiral galaxies passing through the center of the cluster could not retain their disk gas. In clusters of lower total mass and lower galaxy density, and in the outer regions of rich clusters, both the density of the intracluster medium and the expected galaxy velocities with respect to the medium are too low to strip away the disk gas. In such regions spirals are therefore free to complete their normal evolutionary development.

The various models of galaxy evolution that attempt to explain the diver-

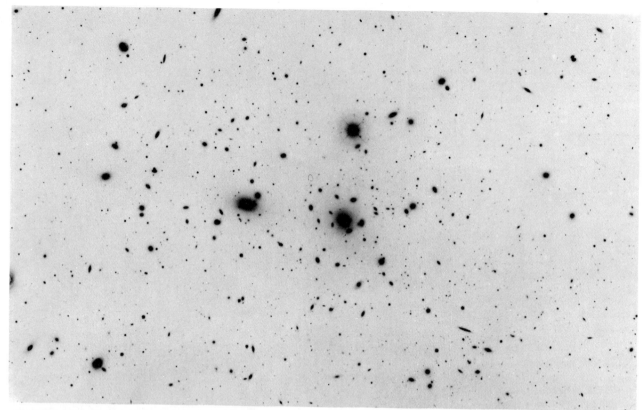

CLUSTER OF GALAXIES IN COMA BERENICES, about 420 million light-years away, is a rich, centrally concentrated cluster with several thousand members, only a fraction of which are included in this negative print of a photograph made with four-meter reflector at Kitt Peak. Central region of the cluster is dominated by SO and elliptical galaxies. As in other rich clusters, spiral galaxies are rare.

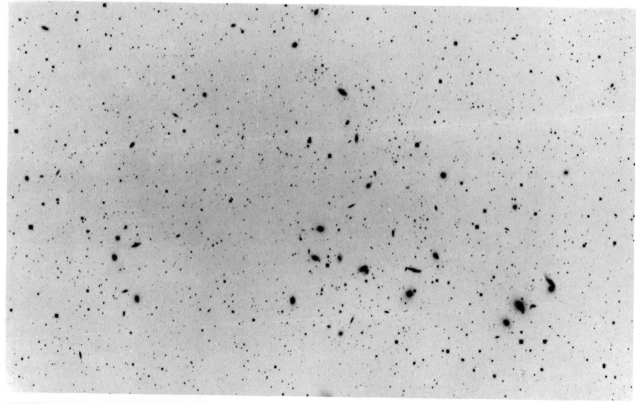

CLUSTER OF GALAXIES IN HERCULES is an open, irregular system 740 million light-years distant. Its several thousand members are packed much less densely than those in the Coma cluster. Unlike the Coma cluster, it includes a high proportion of spiral galaxies. Whereas hot intragalactic gas in the Coma cluster is a strong source of X rays, the Hercules cluster shows little evidence of gas between galaxies and is not an X-ray emitter. The photograph of the cluster was also made with the four-meter reflector at Kitt Peak.

gent pathways that yield either elliptical or spiral galaxies can be tested in three general ways. The first way is to study samples of galaxies in nearby clusters. The second is to compare the characteristics of nearby galaxies with those of the most distant (hence the youngest that can be observed). The third is to search for nearby galaxies that give signs of being younger than typical nearby systems. The three approaches are given in the order of their difficulty, and we shall now discuss them in the same order.

Studies of the detailed distribution of spiral and SO galaxies in nearby clusters of galaxies offer considerable support for the Gott-Gunn explanation of how spirals in rich clusters can be stripped of their disk gas. Jorge Melnick and Wallace L. W. Sargent of Cal Tech and Neta A. Bahcall of Princeton have computed the frequency of occurrence of SO galaxies in a number of galaxy clusters known to be X-ray sources. Their data show that the frequency is highest in clusters that have the largest spread in galaxy velocities and the highest X-ray luminosities. In such an environment the stripping of spirals is expected to be the most efficient. Melnick and Sargent also found that the frequency of SO galaxies decreases in the outer regions of all the clusters in their sample, a result that is consistent with the lower probability of stripping in the cluster regions characterized by a lower density of gas.

At the Kitt Peak National Observatory, working in collaboration with Susan Wilkerson and Eric Jensen, we have recently found examples of galaxies that may have been stripped of their disk gas within the past few billion years. Examination of photographs of X-ray-emitting clusters of galaxies reveals a number of disk galaxies where no newly formed stellar complexes are visible but where the spiral-wave pattern in the old disk stars is still evident. Several examples of such "smooth arm" spirals can be seen in the clusters Abell 262 and Abell 1367 [see illustrations on next two pages]. The amplitude of the spiral wave (as measured by the contrast in surface brightness between the arm regions and the regions between them) seems to be correlated with the measured color of the disk: the highest wave amplitudes are associated with the bluest disks, and the systems with the lowest wave amplitudes have the reddest disks. Blue colors are characteristic of stellar aggregates in which stars have formed fairly recently; red colors suggest that no stars have formed for several billion years.

We believe spirals with smooth arms and a low wave amplitude are galaxies that were stripped of their disk gas and stopped forming stars several billion years ago; they are nearing the end

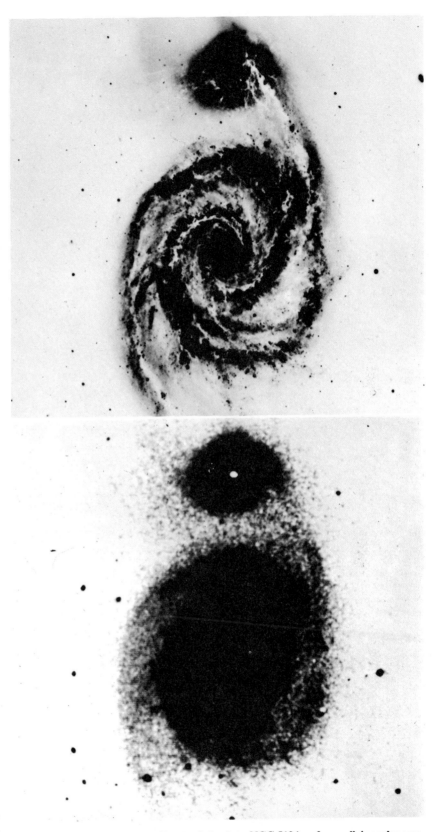

MESSIER 51 consists of a magnificent spiral galaxy, NGC 5194, and a small, irregular companion, NGC 5195, about 10 million light-years away. When the galaxies are photographed in blue light (top), the spiral arms of the large system are clearly delineated by newly formed complexes of hot, young stars and their associated regions of ionized hydrogen. When the galaxies are photographed in the near infrared (bottom), the images are dominated by the population of stars that are older, cooler and redder. The infrared picture reveals the rather smooth wave crests that map the current location of the wave pattern in the disk of the large spiral system. The blue-light picture was made with the four-meter reflector at Kitt Peak. The infrared one was made by Eric Jensen with an image intensifier on the 90-centimeter telescope at Kitt Peak.

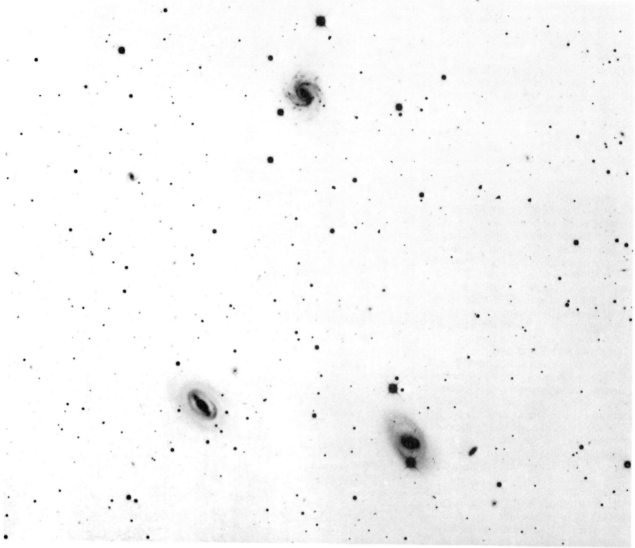

TWO "SMOOTH ARM" SPIRAL GALAXIES, UGC 01350 and UGC 01344, can be seen at the bottom of this negative print of the cluster of galaxies Abell 262. Their appearance is distinctly different from that of the spiral galaxy UGC 01347 (*top center*), in which knots of newly formed stars and ionized gas are clearly visible. It is believed the smooth-arm systems are spiral galaxies whose disk gas has been stripped away from them within the past billion years or so either by an intragalactic "wind" or as a result of the motion of the galaxy through the intergalactic medium. Abell 262, which is about 300 million light-years away, is a cluster that emits X rays.

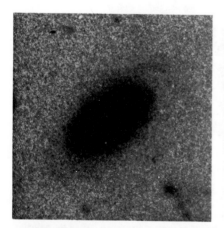 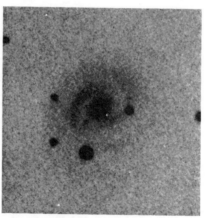 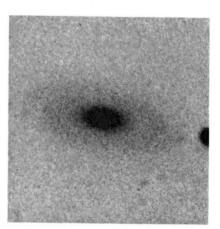

THREE SMOOTH-ARM SPIRALS may represent "snapshots" of spiral galaxies caught in the process of becoming S0 galaxies as their disk gas is stripped away. The galaxy NGC 3860 (*left*), in the galactic cluster Abell 1367, has smooth but clearly visible arms with high-amplitude waves. In NGC 1268 (*middle*), a member of the rich Perseus cluster, wave crests in the spiral arms are much less visible. In IC 2951 (*right*), and also in Abell 1367, the arms have almost disappeared. Analysis of disk colors suggests that NGC 3860 was stripped of gas between 10^8 and 10^9 years ago, whereas IC 2951 was stripped more than 2×10^9 years ago. Photographs were made at Kitt Peak.

of their transformation from the spiral configuration into the SO one. In such galaxies the stellar population of the disk is dominated by longer-lived, redder stars. Spirals with a high wave amplitude and smooth arms may be galaxies that were stripped of their disk gas only a few hundred million years ago. As a result their disk color still reflects the contribution of young, blue stars that were formed just before the gas was stripped away.

The smooth-arm spiral galaxy NGC 3860, located in the X-ray cluster Abell 1367, appears to be surrounded by "shreds" that may consist of ionized gas recently removed from the system. Since no other galaxy in the cluster is surrounded by shreds of similar appearance, the genetic association between the shreds and NGC 3860 seems highly plausible. More definitive proof that the shreds have been stripped from NGC 3860 waits, however, for spectroscopic study. If the line-of-sight velocities of both the shreds and the galaxy are close to the mean velocity of other galaxies in the cluster, one can be almost certain that the shreds are a part of the galaxy cluster Abell 1367. Furthermore, if the abundances of the chemical elements in the shreds (as determined by spectrographic analysis) are similar to the abundances characteristic of the disk gas in normal spiral galaxies, one can conclude that the shreds were torn from the disk of NGC 3860. If the measured velocities and the inferred chemical composition of the shreds indicate they originated in NGC 3860, the stripping hypothesis will have passed its most direct test.

The observed colors of SO systems in the Coma Berenices cluster of galaxies also lend support to the Gott-Gunn picture. When we analyzed the frequency distribution of disk colors for edge-on SO galaxies in the Coma cluster, we found that the outer region of the cluster appears to contain significantly more blue SO's than the central region. The red disk colors characteristic of the central region of the Coma cluster suggest that spiral galaxies were transformed into SO's more than several billion years ago. The presence of blue SO's in the outer region of the cluster can be explained by two plausible hypotheses. The first hypothesis supposes the blue SO's were stripped of their disk gas only recently after an excursion through the central regions of the cluster, where the intracluster gas density is high enough to accomplish the stripping. The second hypothesis assumes that the blue SO galaxies evolved from spiral galaxies in which the gas was not stripped away but was exhausted in the formation of many generations of stars. If the last stars formed within a billion years of the current epoch, the blue relics of these events could account for the observed colors.

The sizes of the disk galaxies in the Coma cluster also seem to show the effect of environment on the evolution of such galaxies. A plot of the frequency distribution of the measured sizes of edge-on SO galaxies in the central and outer regions of the Coma cluster shows that many more large disk systems are present in the outer regions. This result could be explained if star formation were truncated by the removal of disk gas at a fairly early stage in the evolutionary history of the disk systems in the central region. Since the conversion of gas into stars proceeds slowest in the outer region of a galaxy, one can speculate that relatively few stars were formed before the gas was stripped away. SO's in the outer regions of the cluster, and thus less affected by stripping, may have completed a larger fraction of their natural evolutionary history and may therefore have formed a larger number of stars in their outer disk. This interpretation may, however, be complicated somewhat by the possible importance of tidal interactions of galaxies in the dense central region of the Coma cluster. Nevertheless, studies of disk galaxies in nearby clusters tend to support the view that the galaxies' evolutionary history is significantly affected by environmental factors.

The second general approach to testing models of galaxy evolution involves the study of the most remote galaxies. Galaxies with a brightness between 10^{11} and 10^{12} times that of the sun can be observed to vast distances. The time it takes light to travel from the most distant systems observed to date approaches 10 billion years. Observations of remote galaxies located at a variety of distances, or "look back" times, offer the possibility of measuring evolutionary changes directly. Unfortunately galaxies more distant than four billion light-years have an angular size of only a few seconds of arc, which is only slightly larger than the average image distortions introduced by the passage of galactic light through the earth's turbulent atmosphere. As a result even the

FIELD NEAR THE SMOOTH-ARM SPIRAL GALAXY NGC 3860 in the cluster Abell 1367 appears in a color composite photograph prepared with the aid of the Kitt Peak interactive picture-processing system. The colors were obtained by weighting the values in three black-and-white photographs made with the four-meter telescope in the ultraviolet, blue-green and red regions of the spectrum. The yellow-red colors of the spiral are typical of older stars in disk systems. The blue, irregularly shaped shreds may be ionized gas recently stripped from NGC 3860 as i sped through the intracluster gas at some 1,500 kilometers per second.

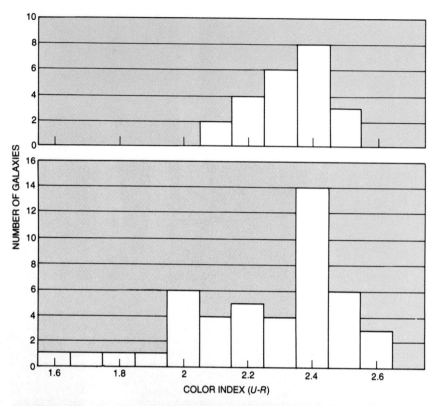

DIFFERENCES IN DISK COLORS are observed when edge-on SO galaxies within 2.1 million light-years of the core of the Coma cluster (*top*) are compared with edge-on SO galaxies more remote from the core (*bottom*). For this comparison the index of disk color (*U–R*) measures the ratio, on a logarithmic scale, of luminosities in the red and near ultraviolet. Large values of *U–R* mean the disk is dominantly "red" and so contains few newly formed blue stars.

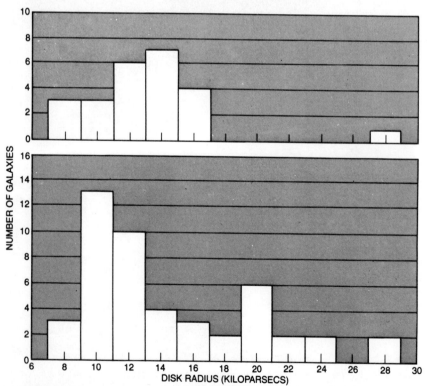

DIFFERENCES IN DISK SIZES are also apparent when edge-on SO galaxies within 2.1 million light-years of the core of the Coma cluster (*top*) are compared with similar but more distant galaxies (*bottom*). It is evident that there are more disks of large size in the outer region of the Coma cluster than there are in the inner region. One kiloparsec is equal to 3,260 light-years.

largest ground-based telescopes are unable to reveal the structural detail needed to classify galaxies as spirals, SO's or ellipticals.

In late 1983 or early 1984 the National Aeronautics and Space Administration will place a 2.4-meter telescope in orbit by means of the space shuttle. The direct-imaging television camera in the orbiting observatory will be able to resolve features separated by only about .05 second of arc, making it possible to classify galaxies and conduct crude structural studies at look-back times of more than five billion years. Until then considerable progress can be made by comparing luminosities and colors measured for distant galaxies with those observed for nearby systems. If reasonably large samples of galaxies are observed, it should be possible to estimate the evolutionary effects.

The feasibility of making such empirical studies of galaxy evolution has recently been demonstrated by Harvey Butcher of Kitt Peak and Augustus Oemler, Jr., of Yale University. With the aid of a sensitive television-camera detector system developed at Kitt Peak by C. Roger Lynds, Butcher and Oemler have recorded the luminosities and colors of a large sample of galaxies in two remote clusters, one seven billion light-years away and the other 8.8 billion. The nearer cluster, known as Cl 0024 + 1654, is receding at .39 times the velocity of light. The farther one, a cluster surrounding the radio galaxy 3C 295, is receding at .46 times the velocity of light.

These two distant clusters appear to be as rich in galaxies and as centrally concentrated as the Coma cluster is. If they were as highly evolved some eight billion years ago as the relatively nearby Coma cluster (whose light has taken less than half a billion years to reach us), the two distant clusters should resemble the Coma cluster, being poor in spiral galaxies and rich in elliptical and SO systems. Since few, if any, stars have formed recently in the elliptical and SO systems of the Coma cluster, the colors of these galaxies are predominantly red. Butcher and Oemler observed that in contrast the brightest galaxies in the distant clusters include many blue-colored systems, presumably spiral galaxies that are producing young, blue stars. Assuming a correspondence between galaxy color and galaxy type identical with that observed for nearby galaxies, Butcher and Oemler deduced that more than half the galaxies in the distant clusters are spirals and that the rest are ellipticals or SO's. In the Coma cluster fewer than 10 percent of all the galaxies are spirals. Hence although the overall character of the distant clusters resembles that of the nearby, spiral-poor clusters such as the Coma cluster, the distri-

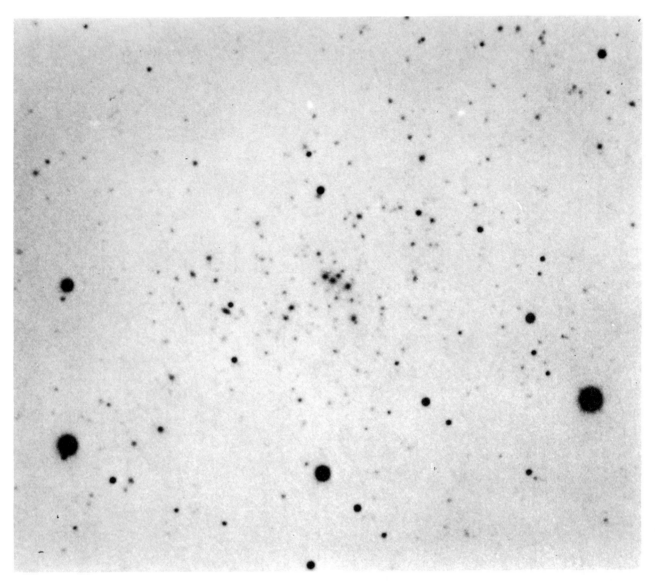

ONE OF MOST DISTANT CLUSTERS, Cl 0024 + 1654, is receding from the earth at .39 times the velocity of light, which implies that it is nearly seven billion light-years away. The study of such clusters can potentially reveal whether galaxies, seen as they appeared long ago, are less highly evolved than the older galaxies that populate much **nearer clusters. The cluster Cl 0024 + 1654 is so far away that its galaxies are barely distinguishable from the images of stars within our galaxy. The cluster consists of several hundred galaxies, most of which are visible in the plate from which this print was made. Photograph was made by C. Roger Lynds with the four-meter Kitt Peak reflector.**

bution of galaxy types in them appears to be quite different.

If the blue galaxies in the two distant, and therefore young, clusters are indeed spirals, it is tempting to suggest that disk galaxies in similar environments have undergone significant evolutionary changes over the past eight billion years. It is possible that in younger clusters of galaxies only a small fraction of spirals have been stripped of their disk gas and transformed into SO's, whereas in Coma-like clusters nearly all the spirals have been stripped. With results available for only two distant clusters it is perhaps premature to embrace this explanation for the differences in the observed color distributions. Nevertheless, the application of sensitive panoramic detectors to the investigation of distant clusters promises to provide additions and challenges to the subject of galaxy evolution.

The third approach to testing hypotheses of galaxy evolution is to search for relatively nearby spiral galaxies whose development appears to be "retarded." Working in collaboration with William Romanishin, we studied a class of spiral galaxies whose surface brightness, or luminosity per unit of area, is so low that they are clearly discernible only on long-exposure plates made with the 1.2-meter Schmidt telescope on Palomar Mountain and with the four-meter Mayall reflector at Kitt Peak. Although such low-surface-brightness (LSB) systems are no more than 60 million light-years away, they are barely

visible above the background sky light. Otherwise they are similar in size and appearance to normal bright spirals. The chief difference is that the disks of LSB galaxies are considerably bluer than the disks of normal spiral galaxies.

The combination of low surface brightness and blue color suggests that compared with the disks of normal spirals the LSB disks are currently populated by relatively few stars. Hence the major contribution to the light of the LSB disks must be made by relatively blue stars younger than three to four billion years. In contrast, the red disk colors of galaxies similar to our own suggest that an overwhelming majority of the disk stars in them were probably formed more than 10 billion years ago. If star formation in LSB galaxies has been relatively inefficient (or possibly absent) until the comparatively recent past, LSB systems may not have consumed a large fraction of the gas initially present in their disks.

Tentative confirmation of this speculation was provided recently by observations of several galaxies of the LSB type at the radio wavelength of 21 centimeters emitted by un-ionized interstellar hydrogen. Working with the 1,000-foot parabolic antenna at Arecibo in Puerto Rico, Nathan Krumm, E. E. Salpeter, Romanishin and we found that some LSB spirals evidently contain more than twice the amount of hydrogen found in normal spirals of a similar type. If LSB galaxies are in fact disk systems that are relatively less evolved than normal spirals, detailed study of this unusual class of nearby spirals may provide a glimpse into the past history of our own galaxy.

The study of galaxies is entering a new phase. In the past much of what is known about galactic structure was derived from the qualitative examination of photographs made with large telescopes. Pioneering efforts to classify galaxies, based on their general appearance, provided the basis for much of the current physical understanding of galactic structure. With the advent of sensitive panoramic detectors and advanced computing methods for the analysis of digitized photographs one can now quantify the color of galaxies and the distribution of surface brightness. For example, many of the new results reported here were obtained with the new interactive picture-processing system (IPPS) developed at Kitt Peak. The system enables one to display digitized pictures of galaxies, to enhance them by emphasizing given brightness levels and to select particular regions for which, say, a brightness or a color measurement is desired. The rapid, real-time interaction with pictorial data increases the efficiency of data analysis, making it possible to study hundreds of galaxies in a variety of settings.

We expect that such quantitative studies will make it possible to address a variety of fundamental questions empirically for the first time. Are galaxies of a given type basically similar in structure or do similar-looking galaxies in rich clusters exhibit measurable differences from their counterparts that are not bound within clusters? If such differences exist, what accounts for them? Can environmental effects such as the stripping of disk gas account for all structural differences, or is it possible that galaxies found in the dense central regions of rich clusters were formed in a way fundamentally different from the way galaxies located outside clusters or in irregular clusters were formed? Is the star-forming history and dynamical evolution of cluster galaxies different from that of noncluster galaxies?

In addition to their implications for understanding the life histories of galaxies, answers to such questions will directly affect the confidence astronomers can place in efforts to use galaxies as standards of brightness and distance in probing the large-scale structure of the universe. One may also learn from such studies the sizes and masses of the first condensations to emerge in the early evolution of the universe. Were these condensations on the scale of stars, star clusters, galaxies or galaxy clusters? Our brief review summarizes only a few of the first results in the renewed quest of understanding how galaxies form and evolve. We hope it conveys some of the excitement of that quest.

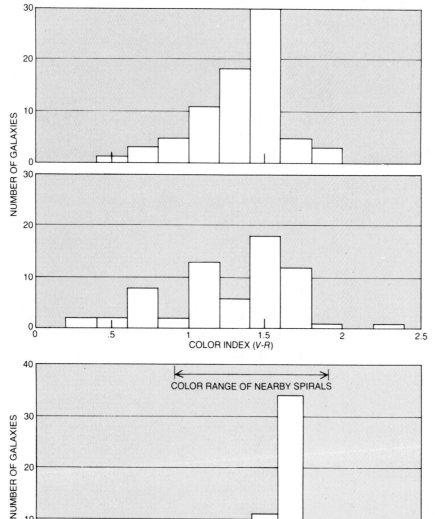

MAJOR DIFFERENCES IN GALAXY COLORS are revealed when the distribution of the colors in the nearby Coma cluster (*bottom*) is compared with the distribution of colors in two very distant clusters, Cl 0024 + 1654 (*top*) and 3C 295 (*middle*). Cluster 3C 295, receding at .46 times the speed of light, is seen as it looked about 8.8 billion years ago. The colors in the Coma cluster are expressed in terms of the index B–V, which measures the ratio of observed "visual" light (wavelength .55 micrometer) to blue light (wavelength .42 micrometer) on a logarithmic scale. The colors in the two distant galaxies are given in terms of the V–R system, in which the red measurement is at a wavelength of .65 micrometer. As plotted, the V–R index is almost equivalent to the B–V index in the rest frame of the receding clusters. The distribution of the colors of galaxies in the distant clusters is distinctly broader than it is in the Coma cluster and in fact is much closer to the range of colors characteristic of nearby spiral galaxies. Thus although the distant clusters are equivalent in richness and concentration of galaxies to the Coma cluster, they appear to contain many more spiral galaxies. This suggests that if the Coma cluster could have been observed as it was seven or eight billion years ago, fewer of its spiral galaxies would yet have been converted into SO and elliptical galaxies by the stripping away of their gas.

Violent Tides
between Galaxies

Alar Toomre and Juri Toomre
December, 1973

*The peculiar forms of certain galaxies could be a result
of tidal distortions. This theory, out of favor for a
decade, seems to be confirmed by experiments
with computer models of galaxies*

Almost every crowd includes a few charming eccentrics or confounded exceptions. This is true of the "crowd" of galaxies. Most are objects of majestic regularity and symmetry and can be readily classified. One or 2 percent, however, do not conform. Because of their bizarre appearance or unusual spectra they are known to astronomers as "peculiar" galaxies.

Many of the galaxies that are peculiar in shape are members of multiple-galaxy systems, and it is only natural to suppose that their unusual and sometimes even grotesque forms may have resulted from the interaction of two or more galaxies. The nature of this interaction has been a matter of controversy, however. In the 1950's, when large numbers of galaxies with strange appendages were first discovered, it was immediately proposed that these morphological anomalies were the aftereffects of gravitational forces exerted during near collisions between galaxies. In the 1960's this idea fell into disrepute, although no alternative theory won general acceptance. In the 1970's computer experiments such as those described here have begun to reaffirm that gravitation may in fact be responsible for the appearance of some of the most peculiar galaxies.

One of the most strikingly peculiar galaxy pairs was discovered before it had been demonstrated that galaxies other than our own exist and can be seen from the earth. In 1917 C. O. Lampland noted that long, faint filaments were visible on improved photographic plates of a double nebulosity listed in the 1888 New General Catalogue (NGC) as entries 4038 and 4039. The objects were photographed again in 1921 with the newly completed 100-inch Hooker telescope on Mount Wilson. J. C. Duncan, who made the Mount Wilson photographs, was particularly impressed by "faint extensions of extraordinary appearance," rather "like antennae" [*see illustration on next page*].

The curving filaments do resemble the antennae of an insect, and the system has become known as the Antennae. It is about 50 million light-years from our galaxy, which among galaxies is not a very great distance. Only about 1,000 easily recognized galaxies lie closer to ours.

The only other notably peculiar object found in this era was discovered by Heber D. Curtis in 1918; he observed a small, luminous "jet" protruding from Messier 87, later identified as an elliptical galaxy at about the same distance as the Antennae. (The Messier catalogue was compiled by Charles-Joseph Messier of France about 1800.)

By 1924 Edwin P. Hubble had deduced that many of the objects then called nebulae were in fact galaxies. The following two decades were an important period of discovery for extragalactic astronomy, yet almost none of the many galaxies identified seemed as bizarre as the Antennae or M87. The only exception, perhaps, was the "faint but definite band of nebulosity" that Philip C. Keenan found to connect two rather widely separated galaxies, NGC 5216 and 5218 [*see top illustration on page 65*].

One reason that scant attention was paid to peculiar galaxies was that an intense effort was being made to understand "ordinary" galaxies. The diversity of forms was great even among those galaxies that clearly fitted Hubble's broad categories of elliptical, spiral and irregular galaxies. In addition, basic questions of the size, distance and velocity of the galaxies remained to be answered. Hubble himself knew of the Antennae and urged Duncan to take an interest in them, but he never systematically searched for similar galaxies.

There is another reason, however, that few peculiar galaxies were discovered in this period: the instruments in use were ill-suited to finding them. As we now know, the "links," "wisps," "plumes," "streamers," "filaments" or "extensions" exhibited by one or two galaxies in a hundred are usually quite faint; they are often little brighter than the night glow of the earth's atmosphere. Such dim, extended features are not easily detected with the type of telescope that was then available. Telescopes such as the 100-inch reflector on Mount Wilson, which has a focal ratio of $f/5$, are quite "slow," that is, they require long exposure times. To photograph the filaments of the Antennae, for example, would take an entire night. In addition, the field of view of such instruments is narrow: about a quarter of a degree, or half the apparent diameter of the moon. For the observation of the faint, extended features of galaxies a "faster" instrument with a wide field of view was needed. A suitable telescope was invented by Bernhard Schmidt in 1931; it provides a focal ratio of about $f/2$ for faster exposures.

A 48-inch Schmidt telescope went into operation on Palomar Mountain in 1949 as a major partner to the just installed 200-inch Hale telescope. For the next seven years this large Schmidt telescope was used to photograph all the northernmost three-quarters of the sky visible

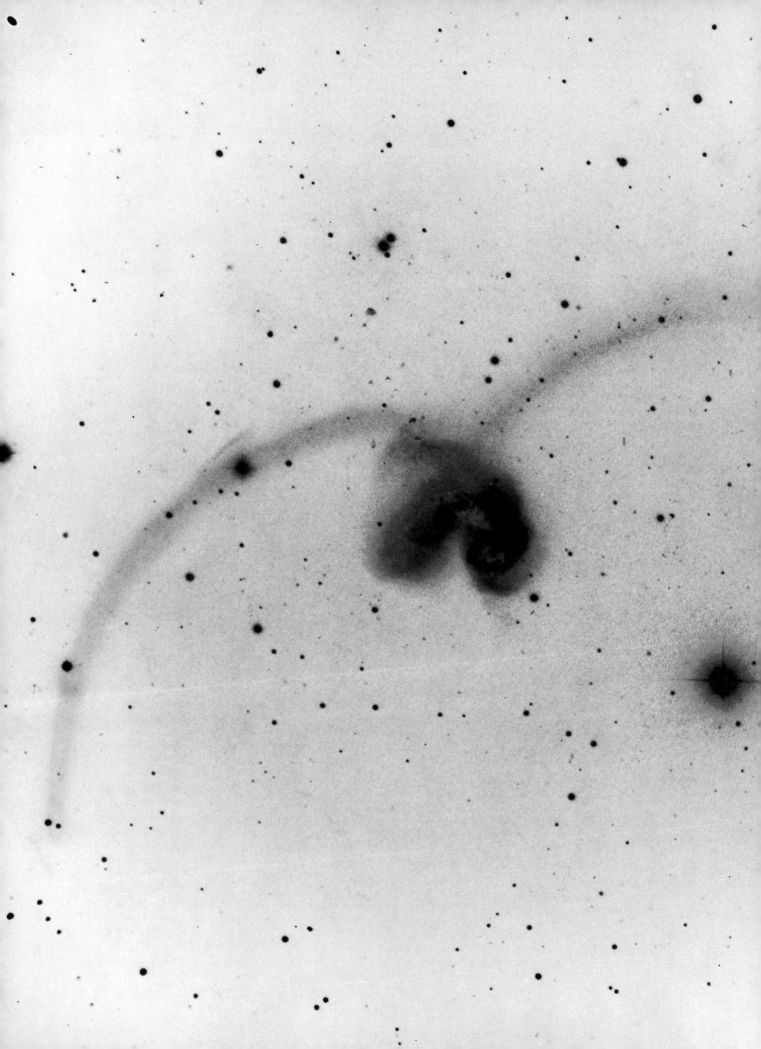

from California in a pattern of almost 900 overlapping square fields, each seven degrees on a side. This was the Palomar Sky Survey; when it was complete, the number of known peculiar galaxies had grown from a few to an untidy and baffling multitude.

By 1956 Fritz Zwicky of the California Institute of Technology wrote that "a surprisingly large number of rather widely separated galaxies appear connected by luminous intergalactic formations." In addition, particularly among close pairs and other multiple galaxies, "many were found to possess long extensions not previously known." Zwicky had somewhat anticipated these discoveries through his work with an 18-inch Schmidt, and he went on to make detailed photographs of many of the galaxies with the 200-inch telescope.

By 1959 the Russian astronomer B. A. Vorontsov-Velyaminov compiled an illustrated catalogue of 355 "interacting" galaxies based on the Sky Survey. Others discovered still more, rephotographed the known ones to larger scale and studied them spectroscopically [see "Peculiar Galaxies," by Margaret and Geoffrey Burbidge; SCIENTIFIC AMERICAN, February, 1961]. Today a particularly fine collection of more than 300 photographs of these objects exists in the *Atlas of Peculiar Galaxies* published in 1966 by Halton C. Arp of the Hale Observatories.

None of these later discoveries closely resemble the Antennae, but a number do exhibit luminous arcs extending far into space from at least one galaxy in a group. All such curving filaments have come to be known as "tails"; two are evident in NGC 4676, a pair of galaxies named the Mice [see *illustrations on page 60*].

The connecting filaments, or "bridges," also assume many forms. One type is represented by the narrow link discovered by Keenan in 1935; Arp's *Atlas* shows many others. One class of bridges seems particularly distinctive. It consists of galactic systems in which a much elongated spiral arm seems almost to grope for a neighboring galaxy. In many cases another limb projects from the opposite side of the deformed galaxy.

Whatever gave rise to these intergalactic bridges and such formations

as the Mice and the Antennae? The answer is not yet certain. It appears, however, that the intuitions of the observers of the 1950's may have been for the most part correct.

The obvious clue available to workers such as Zwicky was that most of the distorted galaxies come in pairs. It was also evident from the shapes and rotational motions of many normally formed galaxies that gravitation remains significant even on a galactic scale, over distances of tens of thousands of light-years. Hence it was natural to wonder if the bizarre forms might represent damage that adjacent galaxies had inflicted on one another by their mere presence and gravity. In other words, could we be viewing colossal tides?

Zwicky believed that tides could in fact explain the tails and bridges. He considered as being particularly good evidence the far-side features, or "counterarms," that usually accompany intergalactic bridges. These offered a crude two-sided symmetry analogous to that of more familiar tides, such as those induced in the terrestrial oceans chiefly by the moon. On the earth high tide comes every 12 hours rather than every 24 because the water level is raised not only at the moving point closest to the moon but also at the point diametrically opposite [see "Tides and the Earth-Moon System," by Peter Goldreich; SCIENTIFIC AMERICAN, April, 1972].

Unlike the orbit of the earth and moon, however, the orbits of most galactic partners probably are not almost circular. For a few years in the 1950's it was widely supposed that many such pairs were not even true satellites or companions. They were believed instead to be mere passersby that had almost collided in their separate courses through space.

One reason this close-approach hypothesis became popular was that only grazing passages seemed capable of producing tides of sufficient magnitude: the tide-raising force varies approximately as the inverse of the cube of the separation of the two bodies, and so it decreases with distance even more rapidly than gravity itself. A second and more enticing reason, however, was that actual collisions between galaxies gave promise of explaining the powerful

sources of radio-frequency signals, such as Cygnus A, that had recently been discovered [see "Colliding Galaxies," by Rudolph Minkowski; SCIENTIFIC AMERICAN, September, 1956].

Cygnus A was the first discrete source of radio waves detected outside the solar system, and it remains the most powerful extragalactic radio source perceived here. It was identified with a visible object by Walter Baade and Rudolph Minkowski in 1951. They found that it is at an enormous distance, by modern estimates about a billion light-years, and that it must therefore radiate prodigious quantities of radio energy. Its optical spectrum was also found to be unusual; it suggests gases in a high state of excitation.

Photographs made with the 200-inch telescope showed two large galaxies that appeared to be almost overlapping. Baade and Minkowski reasoned that the observed radio and optical emissions could be produced if the galaxies were interpenetrating at high speed. In such a collision the probability that the stars of the galaxies will collide is nil, since even within galaxies the distance between stars is vast. Interstellar gases would interact, however, causing mechanical commotion and, it was presumed, copious radiation of the wavelengths detected.

For a few years this explanation of extragalactic radio sources seemed nearly confirmed by observations of two nearer radio sources approximately coincident with the galaxies NGC 1275 and 5128. Both of these sources are very peculiar elliptical galaxies rich in dispersed gases and dust. It was thought once again that their curious features represented other galaxies in transit through them. One can thus understand why Zwicky keenly noted in 1956 that even the Antennae are "a weak radio source and therefore probably a system of two galaxies in the process of a close collision." Minkowski agreed.

Nevertheless, by 1960 the colliding-galaxies theory of radio sources had all but expired. For one thing, the theory proved unable to account for numerous newly discovered sources whose appearance and internal motions gave no hint of any collision. It was also found that the radio emissions of several of the distant bodies come not from the observed galaxies but from regions on each side of them, many thousands of light-years beyond the areas where the impact of gas clouds would be most vigorous. Finally, the radio static was recognized as a type of emission known as synchrotron radia-

LONG, FAINT FILAMENTS curve away from the pair of galaxies NGC 4038 and NGC 4039, also known as the Antennae. These two galaxies, whose "tails" extend across more than a third of a degree of arc in the sky, are almost certainly genuine companions and not merely objects superimposed in our line of sight. The photograph was taken in 1956 by Fritz Zwicky with the 200-inch Hale reflecting telescope on Palomar Mountain. It is printed as a negative rather than in the conventional white on black to accentuate the faint features.

tion (because it resembles the radiation produced by the particle accelerators called synchrotrons). This radiation is emitted by charged particles moving in a magnetic field at speeds close to the speed of light; there is no known mechanism by which the collision of galaxies could yield such particle speeds.

When the collision theory of radio galaxies thus became discredited, tidal effects between galaxies were obviously deprived of an important ally. Left to fend for themselves, they met a barrage of postponed criticism.

It was pointed out, for example, that encounters between unrelated galaxies might be common enough to account for the rare radio sources, but that the space between galaxies is too vast for 1 or 2 percent to have grazed another in a chance meeting. It was also noted that tails emanate from some galaxies that appear to be isolated from their neighbors. Moreover, even if the occurrence of many near collisions could be explained, it was remembered that no one had demonstrated that such encounters could produce the narrow filaments seen in the Antennae, the Mice and certain other distorted galaxies.

In fact, tides seemed plain wrong on two counts. First, Vorontsov-Velyaminov noted that in double galaxies tails are more common than bridges. How can this be reconciled, he asked, with the two-sided symmetry of all known tides? Or even if that symmetry were imperfect, why should the distortion of the far

SEQUENCE OF DRAWINGS by computer shows the very close passage of two identical model galaxies and demonstrates how the encounter between them can produce two tidal tails similar to the tails in the photograph of the Antennae (*see illustration on page 56*). In the computer sequence the mass and gravitational force of each galaxy are concentrated at its center (*large central dot*). Around this central mass rotates a disk of some 350 massless test particles. The two central masses approach each other in elliptical orbits in one plane (*1 and 2*), bringing their disks of particles with them. Even before they reach the instant of closest approach (*3*) they feel a severe pull from the other central mass. Soon the effects due to tidal forces are dominant (*4*), and as more time passes, the material from the far side of each former disk stretches into an ever lengthening tail of debris (*5–7*). The two narrow, arching tails would appear to be crossed, and the bodies of the galaxies would almost overlap if the end product of the encounter (*8*) were viewed from the direction of the arrow. Successive frames are separated by an interval of 100 million years.

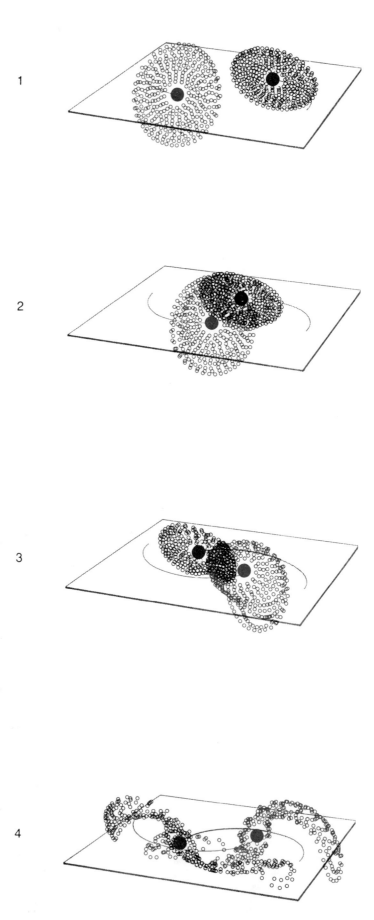

1

2

3

4

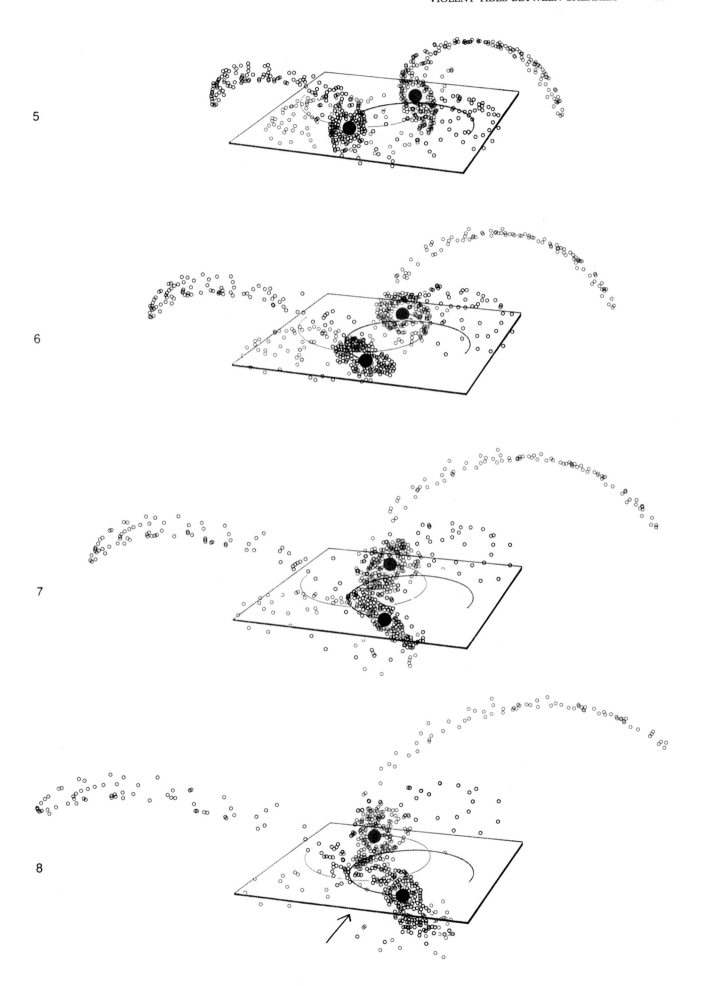

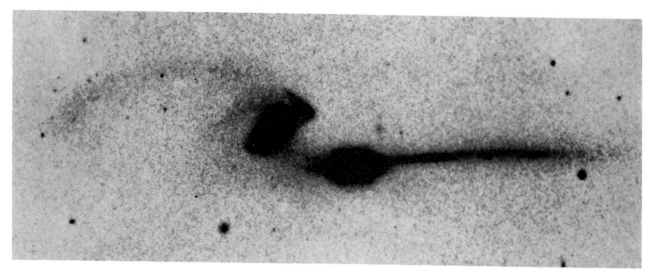

ANOTHER PAIR OF TAILS dominates the pair of galaxies NGC 4676A (*right*) and NGC 4676B (*left*), nicknamed "the Mice." The long tail from NGC 4676A is intense, narrow and almost straight; the one from NGC 4676B appears fainter, more diffuse and much more curved in this photograph made by Halton C. Arp with the 200-inch telescope. The Mice seem roughly one-quarter the angular size of the Antennae yet they are also four times more distant. Hence the true dimensions of the two systems are comparable.

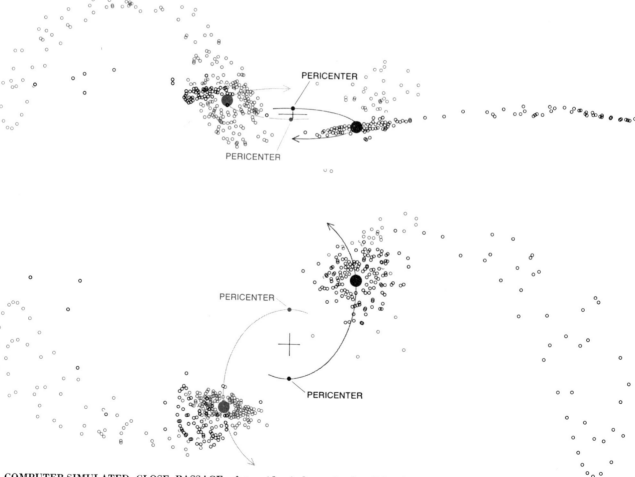

COMPUTER-SIMULATED CLOSE PASSAGE of two identical galaxies yielded the long tails of tidal debris viewed from two separate directions. The top diagram models the actual view of the Mice in the photograph at the top of the page; the bottom diagram repeats the scene as if it were viewed from a direction nearly at right angles. The pericenter is point of closest approach. The central cross marks the center of mass of the system for reference. Long before the encounter each galaxy was again taken to be a circular disk of test particles revolving around a central mass. The two disks grazed each other in elliptical orbits. The tails and surviving bodies of the galaxies appear dissimilar only because each disk was tilted at a different angle with respect to the plane of its orbit around the other. The less tilted disk on the right spilled the material from its far side into a nearly flat tail that lies almost in its orbital plane; the more tilted disk on the left produced the tail that arches high above the plane in the top view.

side be greater than that of the side most exposed to the external gravitational field during an encounter? The other and even more worrisome point was that some bridges and tails are strikingly narrow. This seemed very odd, since known tides raise masses over wide areas. Zwicky himself was troubled by this objection; it seemed to demand that the galaxies behave almost like taffy.

It was not only such intrinsic weaknesses of the tidal theory, however, that caused its widespread abandonment. Also involved was a change of scientific mood or fashion. Even before the first quasars were discovered in the early 1960's, astronomers had developed respect for some highly energetic and even explosive phenomena in galaxies. Supernovae and radio galaxies are but two of many examples. Another is the jet of M87, which, as we have noted, had been known about as long as the Antennae; the discovery in the mid-1950's that even its visible light is synchrotron radiation startled many astronomers. Although poorly understood, all these processes seem to have little to do with gravity. Hence there was much open-mindedness toward invoking other forces, such as electromagnetism, in efforts to interpret various puzzling phenomena.

In this intellectual climate Geoffrey and Margaret Burbidge speculated that the "tubular forms" of such galaxies as the Mice might belong to systems still in the process of formation, probably in the presence of magnetic fields. Vorontsov-Velyaminov wondered if the strange shapes might have resulted from some novel "force of repulsion." More recently it has been suggested that the jet of M87 and some unusual features of other galaxies may have exploded from galactic nuclei. Arp has gone on to ask whether such ejecta might not include some of the small galaxies that seem to be connected to larger ones by bridges.

None of these alternatives of the 1960's were presented as full-fledged theories. They were only hopes or suspicions, no doubt nourished in large part by despair of accounting for objects such as the Mice and the Antennae primarily by gravitation. In this despair were echoes of older remarks, such as "a tidal perturbation can alter the shape of a galaxy but cannot draw out a long narrow filament." Such sentiments had one flaw: it had never been established that gravity could not do it.

Theoretical work since about 1970 has shown that tides can in fact account for some very peculiar structures. As our computer models illustrate, it seems possible after all for a slow near collision to rip the outer parts of a disk into thin and taillike ribbons by gravity alone. It also appears to be possible for other such tides to evolve into remarkably slender bridges and counterarms.

These conclusions are mostly our own, based on extensive experiments we have conducted during the past four years with mathematical models of some double-galaxy systems using the computer and graphic displays of the Goddard Institute for Space Studies in New York. To be sure, like many other overdue ideas, the need for such calculations dawned on several workers more or less simultaneously. Seven workers have recently reported results at least vaguely like ours. One, the Russian physicist N. Tashpulatov, definitely preceded us. Yet the roots of this work go all the way back to an inspired but long forgotten study made by the German astronomer Jorg Pfleiderer about 1960, at a time when chance flybys of galaxies still retained some promise.

Pfleiderer's models, like ours, were intended only as efficient caricatures of the loose confederations of orbiting stars, dust and gas that are the real galaxies. Pfleiderer was no less aware than we are that the mass of an actual galaxy is dispersed over its disk; yet all our models pretend boldly that the matter in their disks is of such small mass that the entire inverse-square gravitational field derives from a single point at each center.

The rationale for using such highly idealized models is twofold. Pfleiderer reasoned that tidal effects should be much the strongest in the exposed and relatively slowly rotating outer parts of the galactic disks. Out there, at least, the mass is small and its self-gravity must be weak. The models should thus remain basically valid even if the peripheral mass is neglected entirely.

Second, as Pfleiderer was particularly conscious in those early days of electronic computers, the idealization of mass as a single point greatly simplifies the computational task of predicting the successive shapes of a disk composed of many particles. For a model galaxy consisting of n particles these numerical economies are roughly n-fold. In the most realistic models possible every particle in a collision of two equal galaxies would be influenced directly by $2n - 1$ other particles; in the simple models the motion of each particle is influenced by only the postulated central masses of the two disks. As it is, the motions of the $2n$ test particles constitute a set of $2n$ dis-

tinct "restricted three-body problems"; such three-body equations have no known analytic solutions of practical value, but they are easily solved in tandem by computer. One can gauge the efficiency of this process by noting, for instance, that our entire simulation of the Antennae (with $n = 350$) could be rerun in less than five minutes on any fast modern computer.

Incidentally, because the test particles are without mass it makes no difference to the motions of one galaxy whether the other arrives with or without its own retinue of massless points. Each of our simulations is thus an anthology of two stories, calculated separately but in the diagrams superposed.

One might well ask why, if these calculations were so inexpensive, we did not adopt more elaborate mathematical models taking some account of self-gravity. The reason is mainly that the construction of even these few examples required hundreds of trial encounters for the purpose of gaining an understanding of the effects of various mass ratios, orbital parameters and times and directions of viewing. These fairly systematic searches revealed that the results are insensitive to changes in certain of the parameters. Yet three conditions seem consistently vital. To produce narrow bridges and tails (1) the galaxies must approach in parabolic orbits or in even slower, elliptical orbits; (2) they must penetrate each other, but not too deeply; (3) the approach of the "attacking" galaxy must be in roughly the same direction as that in which the "victim" disk rotates. Bridges result if the passing galaxy is of fairly small mass, whereas tails require that the two galaxies be nearly equal.

The above may be the crucial ingredients for making bridges and tails, but why do tides assume such forms at all? One reason is that the galaxies themselves are already spinning; the other and less obvious reason is that they experience the intense tidal force only over a relatively short interval. If the sequences portrayed in these computer models actually occurred, they would have required hundreds of millions of years; to the galaxies involved, however, these encounters would have seemed fairly sudden.

The distorting force is not the gravitational field itself but the difference between the fields perceived in the near and far parts of the galaxy. This is in fact the tidal force; as we have noted, it varies inversely not as the square of the distance but as the cube, and therefore it does not become significant until the galaxies are really close. In these computer models it is strong only during the three time frames that bracket the instant of pericenter.

Because of this "hit and run" nature of the tidal force, by the time the tidal damage looks impressive the model galaxies have almost ceased to interact.

Their further evolution is merely kinematic; they drift on independently, like two armadas of spacecraft coasting after a brief, simultaneous firing of engines. Hence the spiral forms (and even the tails) develop not because the imposed gravitational field had a spiral structure but because particles assigned "low" orbits always tend to overtake those in "high" orbits. Those nearer the galactic

nucleus simply shear more and more out of alignment with those farther away.

This qualitative reasoning goes a long way toward explaining the rather two-sided bridge and counterarm that become evident in our computer simulation of the passage of the small companion. The question remains, however, of why the tidal damage in the simulation of the Antennae should have become so much

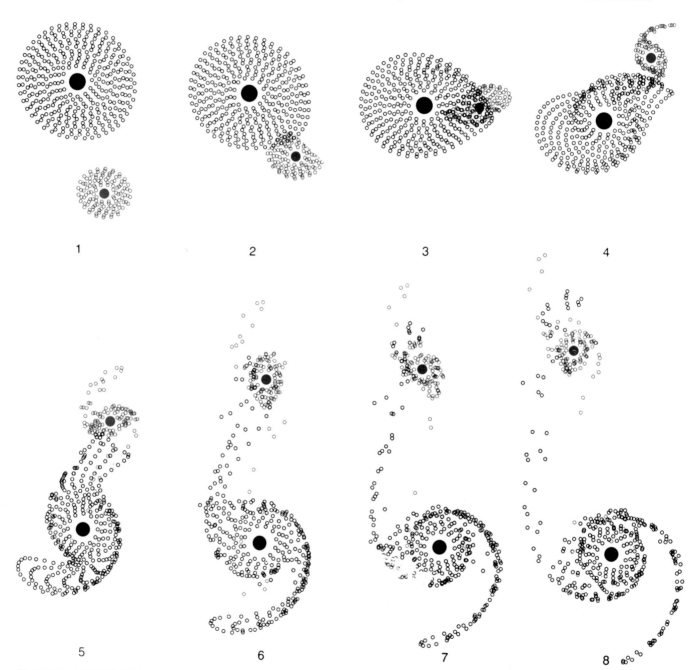

FLYBY OF A SMALL GALAXY produces the striking, if transient, spiral structure seen in this sequence of eight time frames of equal intervals from another computer simulation. Here the body at the center of the larger disk (*black*) is four times more massive than the one at the center of the smaller disk (*color*). Both disks are viewed face on and spin counterclockwise; they encounter each other in parabolic orbits. The plane of their orbits is inclined 45 degrees from the vertical and the disks do not actually interpenetrate. In time frames *1* and

2 the barely distorted small galaxy is still rising toward the viewer. At its closest approach to the larger galaxy (*3*) it passes as much in front of it as to the right of it. The tidal effects in both disks (*4*) are distinctly two-sided. As the smaller galaxy recedes (*5-7*) the tide it raised on the side of the larger disk closer to it evolves into a narrow bridge connecting the two galaxies. The similar bulge that it caused on the far side wraps into a fine counterarm that will become sparse and eventually disappear.

more pronounced on the far side of each galaxy. Part of the answer is that such equal partners obviously damage each other more than the ones in the bridge-building sequence. In fact the near-side tidal forces in that encounter are so great that the material pulled out of one galaxy, rather than forming a bridge, falls in an avalanche into an amorphous mass in the general vicinity of the other. At the same time the far-side material is "launched" vigorously, if indirectly, by having its parent mass practically yanked out from under it. Much of this debris eventually escapes from the influence of both galaxies, resulting in counterarms that grow ever more grotesque and soon dominate the appearance of the galaxy pair. Evidently they are the tails that puzzled Vorontsov-Velyaminov and others.

The reason the orbits must be parabolic or slower is that otherwise the bridges would not connect, nor would there be any avalanching. These failings, if they can be called that, were illustrated by Pfleiderer's calculations with fast but massive passersby. He obtained some fine transient spirals but no bridges or tails.

Since life is too short to watch galaxies move, one cannot be sure that real galaxy pairs orbit each other in the parabolas or elongated ellipses demanded by our models. It should be noted, however, that the statistical objections previously voiced against chance hyperbolic encounters do not apply here. Galaxy pairs in highly eccentric elliptical orbits would necessarily spend a large fraction of their orbital period near maximum separation and would descend only occasionally, like comets, for brief but spectacular displays. At any given moment we would expect to see most such partners well separated from each other, and, unless we knew otherwise, we might never suspect that they were destined to come close. There are in fact many such loose double systems in the sky.

In this discussion devoted to the bizarre it may seem odd to mention M51, a pair of galaxies dominated by the well-known Whirlpool. On most photographs the form and regularity of the Whirlpool show it to be an almost idealized specimen of the spiral galaxy [see top illustration on this page]. It has probably appeared in more textbooks, articles and even advertisements than any other galaxy. Indeed, it was the first galaxy in which a spiral structure was detected (by the Earl of Rosse in 1845).

In spite of the magnificence of its

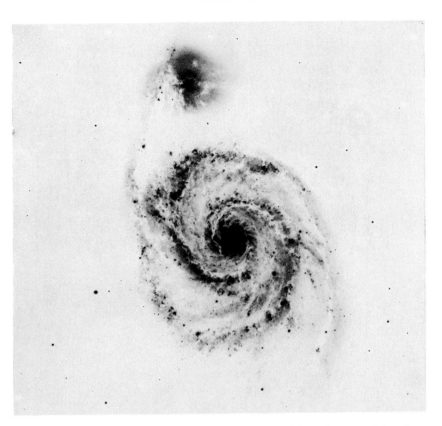

NORMAL PHOTOGRAPH OF THE WHIRLPOOL NEBULA in the constellation Canes Venatici, also known as Messier 51, exemplifies the interior spiral pattern of the stars, dust and gas. The smaller, irregular galaxy appears to be a genuine companion to the larger one.

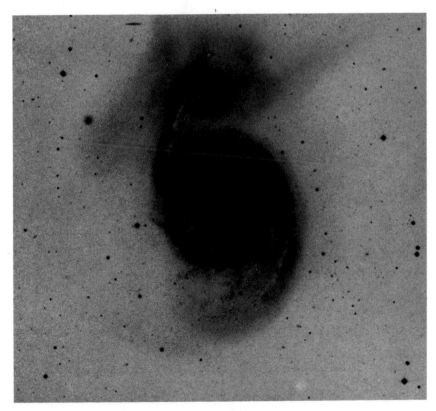

OVEREXPOSED PHOTOGRAPH OF THE WHIRLPOOL NEBULA, taken in 1969 by Sidney van den Bergh with the 48-inch Schmidt telescope on Palomar Mountain, reveals the confusion of faint material that surrounds the smaller galaxy and the two long streamers that extend from it in the directions of two o'clock and eight o'clock. The broad lower arm of the spiral galaxy, prominent in this photograph, is almost invisible in most others.

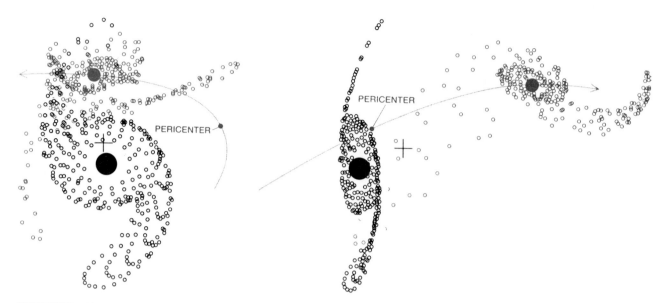

COMPUTER MODEL OF THE WHIRLPOOL NEBULA explores in two views the probable geometry of the encounter that seems to have deformed at least the outer parts of both galaxies. The first view (*left*) shows that the deflected particles from two distinct disks whose central masses are in the ratio of three to one can indeed mimic several of the faint outer features visible in the overexposed photograph at the bottom of the opposite page. The computer model shows both an open-spiral structure in the nearly face-on larger disk and long tidal streamers drawn out from the smaller companion. Second view (*right*) is at right angles to the first and can be thought of as being edge on to the sky; it shows that the arm of the larger galaxy is not a true bridge to the companion.

inner spiral structure, the Whirlpool is a peculiar galaxy. The evidence is abundant in a recent photograph made by Sidney van den Bergh; it shows the faint outer features of both members of the M51 pair about as well as they can be recorded with the telescopes and photographic emulsions available today [*see bottom illustration on page 63*].

Two anomalies of the outer structure revealed by the photograph have been known for some time. They are the "horns" above the companion galaxy and the arm of the Whirlpool that seem to link the two objects. These clues have long suggested tidal damage, but they are inconclusive. Much more significant are the two "plumes" that seem to emanate from the companion. Their importance was first recognized by van den Bergh, although they had been noted two decades ago by Zwicky. Also very interesting is the broad lower extremity of the Whirlpool. This feature, almost invisible in the standard photograph, is most likely a counterarm still in the process of development.

We have calculated the geometry of an encounter between the Whirlpool and its companion that seems qualitatively to explain all those observed outer shapes and yet is consistent with the substantial known speed of recession of the companion [*see illustration above*]. If this hypothetical encounter is realistic,

one fact is clear: the connection between the galaxies is an illusion; the apparent bridge and the more distant companion merely lie in the same line of sight.

Like the vast majority of spiral galaxies, the Whirlpool probably developed most of its fine spiral structure through processes that are intrinsic and have little to do with tides. Yet one wonders: Were the presumed tidal effects such as the outermost arms only superposed on that preexisting structure, or was even the interior of the galaxy somehow rendered more photogenic by the violent tides? These questions remain largely unanswered.

As we have seen, tidal models promise to explain some strange features of galaxies, but they by no means account for all. Of the original threesome of the Antennae, the jet of M87 and Keenan's system that were known to Hubble, only the first now seems to have been plausibly explained by tides. The second appears to be far more esoteric, and as for Keenan's system, no one as yet seems to have any inkling of whether its origin was mundane or exotic.

Many other geometrically peculiar galaxies also remain to be explained. One particularly baffling object is NGC 3921, a galaxy with multiple streamers somewhat like the tails formed in our models [*see bottom illustration on next page*]. Its

deformities are large, yet no second galaxy has been detected anywhere in the vicinity. The hypothesis that we are seeing two galaxies with streamers in the same line of sight might explain this particular object if it were unique, but there are at least five other NGC galaxies in Arp's *Atlas* known to have multiple tails. Even the existence of a "black hole" nearby would not suffice; it might well cause tides, but there is no known way it could cause multiple streamers.

Yet an explanation even of NGC 3921 may not be outside the realm of gravitational dynamics. In our models collisions were assumed to be perfectly elastic; in reality, as the Swedish astronomer Erik B. Holmberg pointed out in the 1930's, collisions between galaxies would involve some frictional forces. It simply costs orbital energy to raise all those violent tides. In the same spirit we suspect that in NGC 3921 and similar objects one is witnessing the vigorous tumbling together or merger of what until recently were two quite separate galaxies.

If this merger hypothesis is confirmed, it will raise the possibility of similar recent goings-on in those peculiar elliptical galaxies that now exhibit either double nuclei or strange interspersed material. It may also provide further impetus to the study of sudden "refuelings" of the very centers of galaxies with fresh interstellar matter.

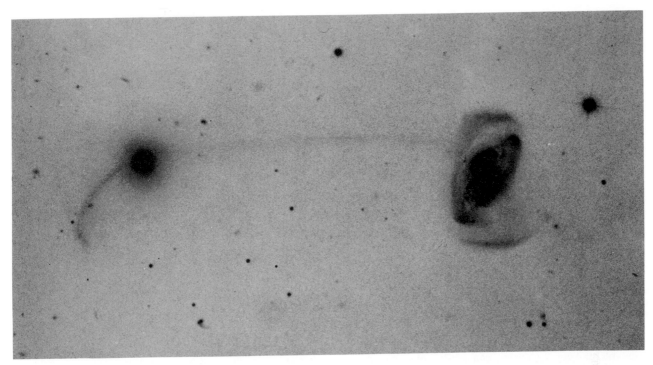

INTERGALACTIC BRIDGE not only connects these two galaxies but also seems to extend through and beyond the one at left. The galaxies, NGC 5216 (*left*) and 5218, are known as Keenan's system after Philip C. Keenan, who discovered the filament in 1935.

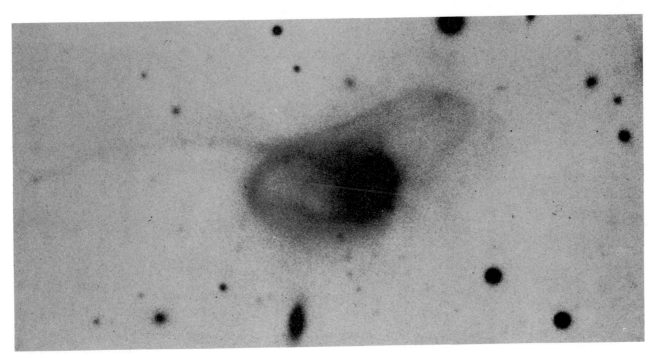

MULTIPLE STREAMERS extend from the object designated NGC 3921. One extends well to the left; another starts in about the same direction but soon turns sharply and seems to end in the faintly luminous region on the right. Authors speculate that two galaxies may here be permanently merging. Both this photograph and the one above were made by Halton C. Arp of the Hale Observatories.

Centaurus A: the Nearest Active Galaxy

by Jack O. Burns and R. Marcus Price
November, 1983

Active galaxies are galaxies that radiate as much as a million times more energy than typical galaxies. The question of what powers the radiation is studied by observations of Centaurus A

Among the largest and most intriguing objects that have yet been observed in the universe are the objects known as active galaxies. Active galaxies, which make up only a few percent of all known galaxies, can emit a million times as much energy in the form of electromagnetic radiation as an ordinary galaxy. The emissions from an active galaxy extend over many different frequencies, including gamma rays, X rays, visible radiation, infrared radiation and radio waves. Much of the recent observational work has been done at radio frequencies, where the emissions from an active galaxy can be 100,000 times as strong as those from an ordinary galaxy.

The active galaxy nearest our own galaxy is Centaurus A, which is about 15 million light-years away. The visible outline of Centaurus A is a few tens of thousands of light-years across. Extending at an angle from the central plane of the visible structure, however, is an elongated radio envelope some three million light-years from tip to tip. If the radio-emitting region of Centaurus A were visible to an observer on the earth, it would appear to be 20 times as wide as the moon.

How is the enormous radio structure of an active galaxy such as Centaurus A created and maintained? The combination of observations made at X-ray, radio and optical frequencies is beginning to tell astronomers a great deal about how Centaurus A is organized. The material of the radio structure is expelled like a stream of water from a hose by an engine at the center of the galaxy. From the engine comes a thin beam of radio plasma: a highly ionized gas whose main components are electrons (moving with nearly the speed of light), magnetic fields and possibly protons. As the electrons travel outward from the center of the galaxy they emit radiation with a wide range of frequencies. In doing so they lose energy, and their motion

would soon stop if they were not continuously supplied with energy. As the beam moves outward from the engine it passes into three radio lobes arranged in order of increasing size. The middle lobe coincides with a series of optical filaments, thin regions that give off radiation in the visible range. The filaments have recently been shown to be regions where new stars are being born.

Although much has been learned about active galaxies in the past decade, some of their most fundamental features are not understood. The nature of the engine that drives the plasma beam is not known with certainty and the details of the mechanism can only be guessed at. Furthermore, there are several possible explanations of how the electrons in the beam are accelerated as they move outward from the galactic core. We cannot choose the correct explanation on the basis of current knowledge. We also do not know how the radio and optical components of the galaxy interact in the regions where the young stars form. Recent observations of Centaurus A at radio and X-ray frequencies have greatly increased our knowledge of what is going on in this active galaxy. In one sense the new information has made Centaurus A more familiar. In another sense the observations have revealed physical processes so large and powerful that they make the active galaxy seem an even stranger and more wonderful phenomenon.

Centaurus A, which in observations at visible wavelengths is designated NGC 5128, was given its label because it is the most luminous source of radio waves in the constellation Centaurus. It occupies a position far south in the sky seen from the Northern Hemisphere. Seen from near Australia in the Southern Hemisphere, the constellation is overhead. Seen from New Mexico, where many of the recent radio observations have been made with the Very

Large Array radio telescope (VLA) of the National Radio Astronomy Observatory, the galaxy is above the horizon only a few hours a day; its maximum elevation is about 14 degrees.

Centaurus A is one of the strongest sources of radio waves in the entire sky. The strength of the emission and the proximity of the object to our galaxy make it possible to observe processes and resolve structures in it that may be present in all active galaxies but that cannot be detected in more distant ones. As a result Centaurus A has had much attention from astronomers in the past few years. It has been studied with many of the most sensitive detectors of astronomical radiation, including in addition to the VLA the four-meter optical telescopes at the Cerro Tololo Inter-American Observatory in Chile and the Anglo-Australian Telescope at Siding Spring in Australia and the Einstein X-ray satellite observatory.

Much of the radiation detected in the work on Centaurus A originates in one of three physical processes. In all three processes radiation is emitted when an electron loses energy. The means by which the electron's energy changes, however, is different in each process, and the characteristics of the emitted radiation also differ. In the first process clouds of interstellar gas are energized by incident radiation. In such clouds atoms of gas absorb radiant energy and are thereby ionized (meaning that they gain or lose electrons). When an electron in the cloud comes near such an ion, the electron is deflected and accelerated; most of the ions are protons, hydrogen nuclei stripped of their single electron. As the electron's path shifts it emits a photon, or quantum of electromagnetic energy. Radiation generated in this way is referred to as thermal radiation because its spectrum depends strongly on the temperature of the gas.

Electromagnetic radiation can be described either as a stream of photons or as a train of waves. In the photon description the radiation can be characterized by the energy of the photons; the corresponding property in the wave description is the frequency, which in turn is directly related to the wavelength. The greater the energy of the photon, the higher the frequency of the waves and the shorter the wavelength. In thermal radiation the energies of the emitted photons or the frequencies of the emitted waves are distributed continuously, that is, they form a smooth curve when they are plotted on a graph. The radiation tends to be most intense, however, in a fairly small range of frequencies. The waves are randomly polarized, meaning that the plane of the electromagnetic oscillations can have any orientation.

In the second process an electron is captured and accelerated in a helical path around a line of force in a magnetic field. As the electron is accelerated it

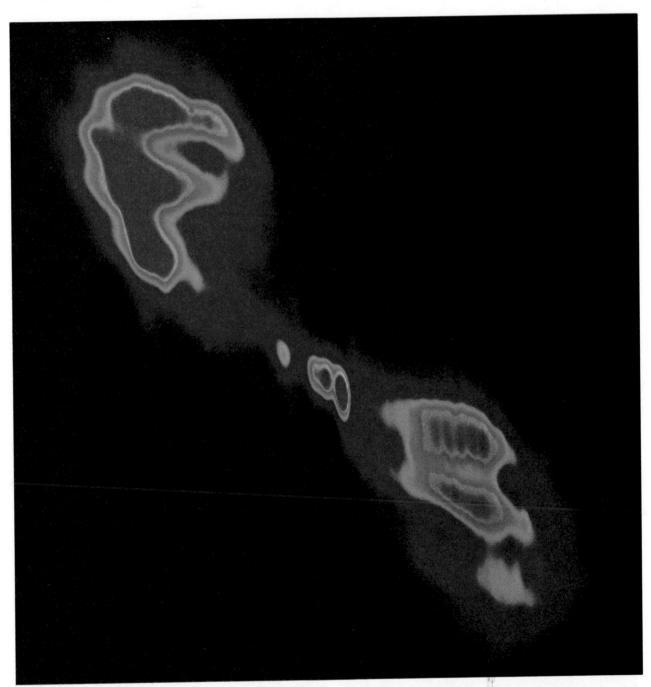

INNER RADIO LOBES of Centaurus A are shown on a false-color map of the radio emissions from the central region of the galaxy, which at visible wavelengths is designated NGC 5128. In the arbitrary color scheme red stands for the most intense radio emissions and blue for the least intense. Here and in other images of the galaxy on the following pages north is at the top and east is at the left. The inner lobes are the symmetrical rounded structures aligned on a northeast-southwest axis. The distance from tip to tip is about 60,000 light-years. The inner lobes are the smallest and innermost of a series of radio structures that extend from the center of the galaxy on each side. The narrow channel between the center of the galaxy and the northern inner lobe is called a jet. It is thought to be a beam of plasma, a highly ionized gas. The plasma is ejected from the galactic core, which is marked by the larger of the red ovals between the inner lobes. As in other active galaxies, the emissions from the plasma have a broad range of frequencies. The radio emissions were observed by one of the authors (Burns), Eric D. Feigelson and Ethan J. Schreier with the Very Large Array radio telescope (VLA) in New Mexico. The map was made by means of a computer image-processing system developed by Dennis Ghiglia of Sandia National Laboratories.

emits electromagnetic radiation tangent to the helical path. The emission constitutes what is called synchrotron radiation. Synchrotron radiation, like thermal radiation, has a continuous frequency distribution, but it is spread out over a broader range of frequencies than thermal radiation. Moreover, in synchrotron radiation the waves are polarized: the electric-field components of all the waves are aligned in the plane perpendicular to the magnetic line of force around which the electron spins.

In the third process a photon is emitted when an electron moves from one of the allowed energy levels in an atom to a lower level. When an atom in a cloud of gas absorbs energy from its environment, an electron in the atom can move to a higher energy level. Having ascended, the electron tends to quickly return to the original energy level. In the downward transition a photon with an energy equal to that of the photon absorbed in the upward transition is released. The photons emitted in this way have a set of discrete frequencies, each frequency corresponding to a particular transition within the atom. On a continuum representing the total electromagnetic spectrum the emitted radiation appears as a group of the bright lines known as atom-

ic emission lines. The radiation in the emission lines is randomly polarized.

As we shall see, synchrotron radiation has turned out to be of great significance in understanding the physics of Centaurus A. The earliest observations of the galaxy, however, were made by means of radiation in the visible range, which is not surprising since NGC 5128 is the seventh most luminous object in the sky outside our own galaxy. From a wealth of optical observations beginning early in the 19th century it has become clear that the visible part of Centaurus A has a very unusual structure, with features of both elliptical and spiral galaxies.

OPTICAL IMAGE AND RADIO CONTOURS of Centaurus A are aligned perpendicular to each other. The illustration combines the radio contours of the inner lobes with a picture of the galaxy made with radiation in the visible range. The visible area of NGC 5128 has an unusual form with features of both elliptical and spiral galaxies. The large luminous circle is much like the body of an ellip-tical galaxy. The dark stripe that bisects the circle is a disk-shaped dust lane composed of gas, dust and associations of stars; the dust lane resembles a spiral galaxy. The disk is warped in opposite directions at its ends. Although the ellipse is rotating slowly if at all, the disk of the dust lane is rotating steadily. The radio structure emerges from the center of the dust lane more or less along its axis of rotation.

The main body of NGC 5128 is a bright ellipse about seven minutes of arc in angular extent. At the estimated distance of 15 million light-years each minute of arc is approximately 5,000 light-years, and hence the ellipse is about 35,000 light-years across at its widest. The mass of the galaxy has been estimated at about 300 billion times the mass of the sun. This mass and size are typical of elliptical systems.

What is unusual is the presence of a broad absorption lane, or lane of dust, in the middle of the ellipse. In optical images the dust lane appears as a dark stripe aligned with the short axis of the ellipse. The lane has the shape of a disk with its edges warped in opposite directions, like the brim of a fedora.

The dust lane is populated by dust, associations of stars and regions filled with ionized atomic hydrogen. The neutral, or un-ionized, hydrogen atom is designated H I and the ionized atom is designated H II. Therefore the regions of ionized hydrogen observed in the dust lane of NGC 5128 are referred to as H II regions. In the H II regions the hydrogen gas has a temperature of some 10,000 degrees Kelvin (degrees Celsius above absolute zero). The gas is heated by radiation from hot young blue stars that have recently formed in the gas cloud. The H II regions give off a set of characteristic atomic emission lines.

The main body of Centaurus A, like that of many elliptical systems, is rotating very slowly if at all. The dust lane, however, is rotating steadily and in a pattern typical of spiral or disk galaxies. The dust lane and the ellipse are centered at the same point, and the gravitational potential of the large ellipse dominates the system composed of the two components. Such a hybrid configuration is extremely rare and could well be connected with the violent activity underlying the radio structure.

Before turning to the radio and X-ray observations that are the center of current interest in Centaurus A, one other unusual optical feature is worth noting. Long-exposure photographic plates made by J. A. Graham of Cerro Tololo show there is a remarkable elongated luminous region in the northeast quadrant of the galaxy. The plates were made with the light of the atomic emission line corresponding to the recombination of an electron with a proton in the H II regions, which is referred to as the H-alpha transition. The plates show that at a distance of 50,000 light-years from the center of the galaxy there is a linear filament including three large H II regions aligned radially outward from the center of the galaxy. Farther out, at distances of up to 130,000 light-years, are regions yielding emission lines and including both filaments and diffuse gas. Spectroscopic studies show that the

excitation of the gas decreases with distance from the galactic core, suggesting the source of the excited gas is in the center of the galaxy.

Near some of the elongated H II regions are what appear to be chains of bright blue young stars; the chains were first noted by Victor M. Blanco and his colleagues at Cerro Tololo. If these are indeed stars less than 10 million years old, they must have formed near where they are now observed rather than forming at the center of the galaxy and then being ejected. The stars are about 65,000 light-years from the galactic core, and even at the unusually high velocity of 1,000 kilometers per second (which is many times the velocity needed to escape the gravitational field of the main body of the galaxy) the stars could not have existed long enough to travel from the center of the galaxy to where they now are.

The first detailed radio observations of Centaurus A were begun in 1961 when the 210-foot radio telescope at the Australian National Radio Astronomy Observatory came into service. One of us (Price) was fortunate enough to work as a graduate student with J. G. Bolton and B. F. C. Cooper of that observatory in utilizing the new telescope, which was then the premier instrument in radio astronomy, to observe Centaurus A. It was quickly found that the radio emission from the galaxy is plane-polarized, suggesting it comes from the synchrotron process.

Further work showed that the plane of polarization of the emission is rotated as the radio waves pass through clouds of cosmic-ray electrons and the weak magnetic fields in the interstellar space of our galaxy. Such rotation, which is called the Faraday effect, is proportional to the square of the frequency of the radiation. Although the Faraday effect had been predicted theoretically, the observations of Centaurus A with the Australian radio telescope marked the first time the effect had been clearly detected observationally.

Thus the early radio investigation of Centaurus A had two significant results. First, the existence of the interstellar magnetic fields in our own galaxy was confirmed. Second, the confirmation of the hypothesis that the radio emissions from NGC 5128 come from a synchrotron process was a significant step in understanding the active galaxy.

Furthermore, the Australian radio observations provided the first reliable map of the overall radio structure of Centaurus A. It was found that the linear extent of the radio region is 2.7 million light-years, covering about 10 degrees in the sky. When Centaurus A was first mapped in the early 1960's, it was the largest discrete astronomical object that had ever been observed. In the past

10 years larger radio sources have been found, but Centaurus A is still among the largest astronomical objects known.

The extended radio-emitting regions on each side of the galactic center change their direction with the distance from the nucleus. Near the nucleus the alignment is northeast-southwest but the outer regions are aligned almost north and south. The shift in direction of about 65 degrees from the center of the galaxy to the tip of the radio envelope is caused by a continuous curvature along the length of the radio region. As a result the overall radio structure has the shape of a huge S. The curvature indicates that the flow of material into the radio envelope is being continuously perturbed. The perturbation could be due to a precession (a wobble like that of a top) of the nuclear engine or to the effect of gas in the galaxy outside the nucleus pressing on the plasma beam.

Between the galactic nucleus and the tip of the radio envelope is a series of complex radio structures. The size of these substructures increases from the nucleus outward, and the scale of the largest ones is about a million times that of the smallest. The details of the radio components have been resolved only recently in work with the VLA done by us, Eric D. Feigelson of Pennsylvania State University, Ethan J. Schreier of the Space Telescope Science Institute and George W. Clark of the Massachusetts Institute of Technology.

In a radio interferometer such as the VLA radio signals from a single source are received simultaneously by an array of several antennas. In the VLA there are 27 antennas arranged along three intersecting arms. Comparing the phase relations of the waves arriving at the receivers can give a rich picture of the radio source. With the VLA we have been able to achieve a resolution of one second of arc. At the distance of Centaurus A such a resolution is equal to about 80 light-years and is comparable to that in photographic plates made with visible radiation gathered by large telescopes.

Moving inward toward the core from the ends of the radio structure an asymmetry becomes clear. To the northeast, between 70,000 and 130,000 light-years from the nucleus, is a large, roughly circular middle radio lobe. No comparable middle lobe is seen at the corresponding position in the southwest. The middle lobe is a region of intense emission that is distinct from the diffuse outer radio envelope. The position of the middle lobe coincides with that of the optical filaments where young stars are seen. As we shall see, it also overlaps a newly discovered area of X-ray emission.

Closer to the nucleus is a pair of inner radio lobes, one lobe on each side of the core. The inner lobes, which are also distinct from the outer envelope, are symmetrically positioned 30,000 light-years from the center of the galaxy. With a diameter of 3,500 light-years they are considerably smaller than the middle lobe. No visible radiation or X-ray emission has yet been observed in association with the inner lobes.

Connecting the northern inner lobe and the nucleus is a narrowly collimated channel of radio and X-ray emission called a jet. Such jets, which are observed in many radio galaxies, could be channels that carry freshly energized electrons and magnetic fields from the central engine to the inner lobe. If this is so, the jet is a leaky and inefficient channel, since some of its energy is radiated outward as synchrotron emission.

Perhaps the most striking asymmetry in Centaurus A is the fact that only the northern part of the radio envelope has a jet; no comparable structure is observed in the southern part. Many radio galaxies have a jet extending from the galactic core, and in some of them the jet is one-sided. The reason for the asymmetry is not known, but several hypotheses have been proposed. Lawrence Rudnick of the University of Minnesota has suggested that periodically the jet switches sides. As a result over any long period the lobes would be supplied with roughly the same amount of energy. The oscillation could be due to the interaction of the jet with the dense gas in the region of the nucleus. Other workers have proposed that jets actually extend from both sides of the core but that because of relativistic effects the emission from the jet coming toward us is greatly enhanced and the emission from the jet going away from us is greatly reduced

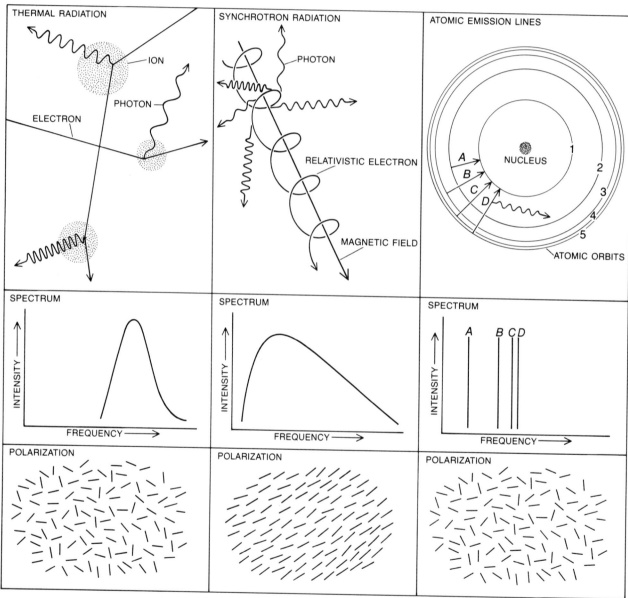

RADIATION EMITTED BY CENTAURUS A could originate in one of three physical processes. In all three an electron loses energy, and the mechanism by which the energy is dissipated determines the frequency and intensity of the radiation. In thermal radiation (*left*) electrons are deflected and accelerated as they pass near ions (mostly hydrogen nuclei, or protons) in a cloud of hot gas. When an electron is accelerated, it gives off a photon, or quantum of radiation, which can be said to have an energy, a frequency and a wavelength. The frequencies of thermal radiation are continuously distributed, but the radiation is most intense in a fairly narrow range. The polarization of thermal radiation is random. In synchrotron radiation (*middle*) photons are given off by relativistic electrons: electrons with a velocity approaching that of light. The relativistic electrons are accelerated in a helical path around lines of force in a magnetic field and photons are emitted tangent to the path. Synchrotron radiation has a broad range of frequencies and is polarized. An electron can also give off radiation when it makes a downward transition from one allowed energy level in an atom to a lower level (*right*). The spectrum of radiation that is emitted in this way consists of a series of discrete lines. The radiation in these atomic emission lines is randomly polarized.

[see "Cosmic Jets," page 77]. This explanation, however, requires that the bulk speed of the plasma be close to that of light, which appears to be unlikely in Centaurus A.

In radio interferometry the resolution of the observations increases when the base line, or the distance between antennas, is increased. The resolution that can be achieved with a radio interferometer has recently been improved by putting antennas on different continents. The data from each instrument are recorded on a magnetic tape and are later correlated and compared by computer. Robert A. Preston and his colleagues at the Jet Propulsion Laboratory of the California Institute of Technology employed radio telescopes in Australia and South Africa to observe Centaurus A and were able to resolve features as small as three milliseconds of arc in the nucleus of the galaxy.

Deep in the core of Centaurus A, Preston and his colleagues found a jet about 50 milliseconds of arc long. At the distance of Centaurus A, 50 milliseconds of arc is equal to four light-years. The small jet emerges at the same angle as the larger radio jet found with the VLA. This remarkable collimation of the beam from four light-years out to 20,000 light-years means that the engine in the core maintains a consistent spatial orientation over long periods. The small inner jet is very close to the engine and is probably the initial point where the plasma beam becomes collimated.

Completing the radio map of the galaxy is the core itself. In radio observations the core appears as a strong source of emissions at the very center of the galaxy; it cannot be resolved into its components with current instruments. While the detailed structure of Centaurus A was being examined by means of radio interferometers another significant kind of information about the galaxy was being accumulated.

The X-ray telescope in the Einstein observatory satellite, launched in November, 1978, was the first instrument capable of detecting and mapping soft X rays from sources other than the sun. Within the X-ray part of the electromagnetic spectrum soft X rays are those with a relatively low energy: from about 1,000 to 10,000 electron volts. Hard X rays correspond to photons with a high energy: from about 10,000 to 50,000 electron volts. The two main detectors mounted on the Einstein satellite were able to resolve X-ray features respectively 1.5 minutes of arc and five seconds of arc across, for the first time giving X-ray astronomers a resolving power comparable to that of radio telescopes and optical telescopes.

Soon after the Einstein satellite was

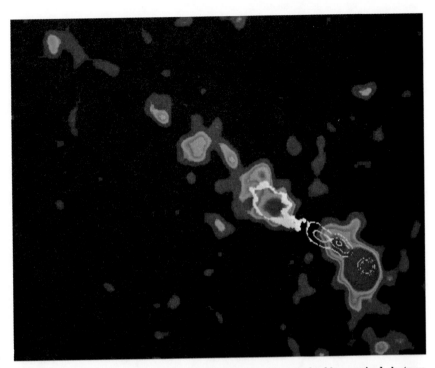

INNER JET is a highly collimated beam of plasma that channels freshly energized electrons from the core of Centaurus A to the northern inner lobe. The galactic nucleus is the compact red region at the lower right with white contours superposed on it; the jet extends to the northeast from the core. The jet is about 3,000 light-years long. The computer image combines radio data with X-ray data gathered by Feigelson and his colleagues using the Einstein satellite observatory. The colored areas correspond to X-ray emission, the white contours to radio emission. The conjunction of X-ray and radio emissions suggests that the radiation from the plasma comes from a synchrotron process. No corresponding jet is seen to the southwest of the core.

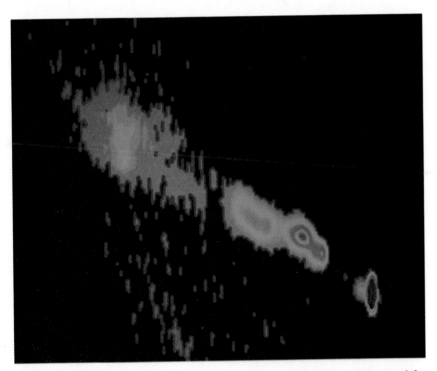

HIGH-RESOLUTION IMAGE of the inner jet is a detail of the illustration at the top of the page: it expands a small area near the center of the galaxy. The galactic core is the red oval at the lower right. The colored areas to the northeast of the core are "knots," regions of particularly intense radio emission. Since the relativistic electrons in the plasma can retain their high energy for only about 50 years, they must be reaccelerated in order to reach the inner lobe, which is 3,000 light-years from the core. The knots could be turbulent regions in the jet where the electrons are reenergized. The image was made utilizing the VLA in a configuration that makes it possible to resolve features as small as 80 light-years across at the distance of Centaurus A.

launched Feigelson and Schreier, together with Riccardo Giacconi of the Center for Astrophysics of the Harvard College Observatory and the Smithsonian Astrophysical Observatory, directed the satellite's instruments at Centaurus A and found a wealth of detail unprece-

dented in extragalactic X-ray astronomy. At about the same distance and angle from the nucleus as the northern middle radio lobe is a region of X-ray emission 24,000 light-years long. The radiation is more intense than would be expected from synchrotron emission.

Therefore it is probably thermal radiation; its source could be a cloud of gas with a temperature of 10 million degrees K. The gas might be heated when it is compressed by turbulence in the radio plasma of the middle lobe.

An envelope of less intense X-ray

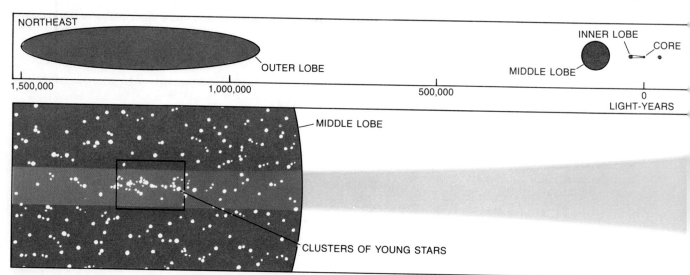

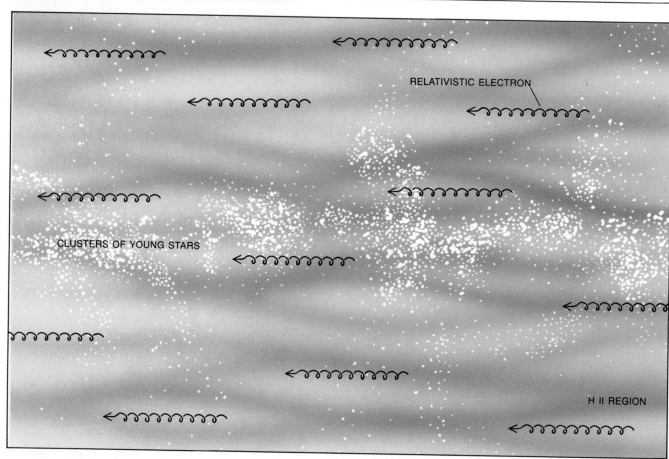

DYNAMICS OF THE EMITTING REGION of Centaurus A are shown schematically in panels with a progressively smaller scale. The full extent of the emitting region is three million light-years (*top panel*). Radio emissions are shown in red and X-ray emissions in blue. The pair of inner lobes on each side of the center of the galaxy are areas of relatively intense radio emission within a diffuse outer emitting envelope. The inner jet that connects the core to the northern

inner lobe gives off much emission at both radio and X-ray wavelengths (*middle panel*). It lies within a larger region of diffuse X-ray emission that comes from the interstellar medium. The region of diffuse X-ray emission extends to the northern middle lobe. One edge of the middle lobe is also the site of a group of extended filaments and bright blue point sources of visible radiation. One hypothesis that accounts for many recent radio and X-ray observations of Centaurus emission surrounds the galaxy, extending out to a distance of about 10,000 light-years from the core. The diffuse emission is probably generated by a cloud of interstellar gas with a temperature of 20 million degrees K. and a total mass 200 million times that of the sun.

The detection of the hot gas by the instruments on board the Einstein satellite is one of the few times an interstellar medium has been found within the visible body of an elliptical galaxy.

With a density of about one particle per 300 cubic centimeters the interstel-

lar medium discovered in the Einstein satellite observations is fairly dense by astronomical standards. The dense medium could confine the plasma in the radio jet and the inner radio lobe and thereby contribute to the maintaining of their shape. Such confinement could be

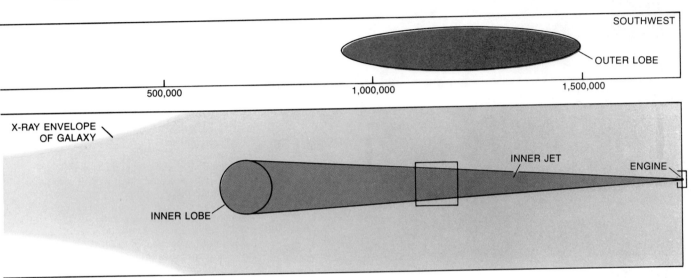

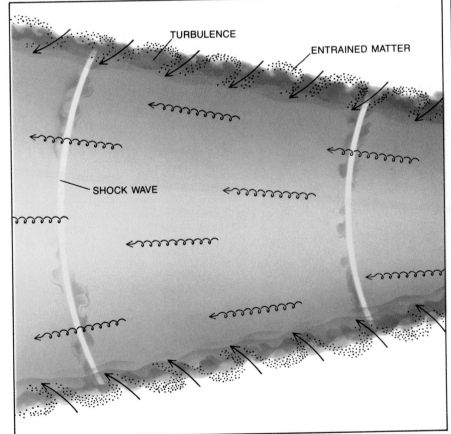

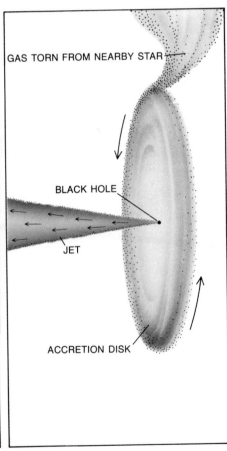

A is the following: The engine that generates the radio plasma has at its center a black hole with the mass of a billion suns (*bottom right*). The black hole is surrounded by a toroidal accretion disk made up of gas and dust. Fuel for the engine comes from infalling gas, which could be tidally torn from the atmosphere of stars near the disk. The interaction of the gas with the black hole and the accretion disk yields a narrowly collimated beam of electrons and magnetic fields: the inner jet. As the jet moves outward from the core, turbulence at the edge of the jet causes gas to be picked up and carried along with the plasma (*bottom middle*). Within the jet shock waves reenergize the relativistic electrons. When the plasma reaches the region of the middle lobe, some of the entrained gas cools and contracts and forms stars (*bottom left*). At the end of their life they explode as supernovas, putting energy back into the radio plasma and energizing the electrons.

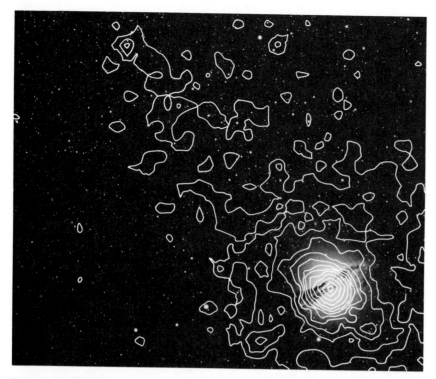

X-RAY CONTOURS mapped with the Einstein satellite observatory are superposed on an image of NGC 5128 made by means of visible radiation. The main body of the galaxy is at the lower right. The optical filaments and young stars near the northern middle lobe are at the upper left; they are 130,000 light-years from the core. The chief X-ray emissions outside the main body of the galaxy coincide with the optical filaments. The optical image is made by the emission line resulting from the recombination of an electron and a proton to form a hydrogen atom.

RADIO CONTOURS made with the 210-foot radio telescope at the Australian Radio Astronomy Observatory are superposed on the same optical image of NGC 5128 in the illustration at the top of the page. As in the case of the X-ray emissions, the region of the most intense radio emission outside the main body of the galaxy (the northern middle radio lobe) coincides with the optical filaments and young stars. The overlapping of the sources of radiation at X-ray, visible and radio frequencies suggests that the same process could be responsible for all three.

critical. Without it the radio jet might lose its collimation and the lobes would expand greatly, diluting their energy density. If the hypothesis is correct, the structure of NGC 5128 confirms the thermal confinement of radio structures, a phenomenon that had been predicted but had never been observed.

The most exciting finding made with the Einstein satellite, however, was the discovery of a jet of X-ray emission coming from the core in much the same position as the VLA radio jet. This was the first time such an intense elongated structure had been found in X-ray data. Indeed, the X-ray jet was discovered before the VLA jet, and the X-ray findings motivated much of the recent work at radio frequencies.

Not only is the X-ray jet the same size and shape as the radio jet but also at the same places in the two jets there are "knots": small regions where the emissions are particularly intense. The structural similarities, along with the continuous spectrum of the emissions and their polarization properties, led us to conclude that the X-ray emission from Centaurus A is also synchrotron radiation.

When the data from more than two years of operating the Einstein satellite were analyzed, only one other galactic X-ray jet like the one in NGC 5128 was found. The second jet is in the next-closest active galaxy: M87. This galaxy is about three times as far from us as Centaurus A is. The resolution of the satellite's instruments, however, is not sufficient to show structural details in the M87 jet. The X-ray jet in Centaurus A has details that can be observed individually, and it therefore currently remains a unique astronomical object.

What does the detailed map of NGC 5128 that has been patiently assembled at X-ray and radio frequencies tell us about the physical processes going on in the galaxy? Consider the inner radio and X-ray jet. From the point of view of astronomical observation the most significant component of the jet is the stream of relativistic electrons (electrons moving at nearly the speed of light). The electrons are the source of the synchrotron radiation. It does not follow, however, that because the electrons are moving at relativistic velocities the jet as a whole has such a velocity. The beam includes a considerable quantity of gas in addition to the electrons, and the gas has a much lower speed than the electrons do. The best current estimate is that the bulk-flow velocity of the jet is about 5,000 kilometers per second.

The conclusion that the X-ray photons detected by the Einstein satellite come from a synchrotron process puts some severe constraints on the model of what goes on in the jet. The X-ray photons have an energy of about 2,000 electron volts. In the synchrotron process such photons are emitted by relativistic electrons with an energy of 3×10^{13} electron volts. Electrons with that much energy cannot stay in a magnetic field for long before they emit photons.

If the emission continues, eventually the electron will be depleted of energy and will not be able to emit any more photons. Indeed, given the energies of the electrons and the X-ray photons the depletion will occur in 50 years or less. The radio and X-ray jet in Centaurus A, however, reaches the inner lobe, which is 20,000 light-years from the nucleus. Therefore if the electrons are traveling at about the speed of light, they will have to be reenergized many times in order to reach the inner lobe.

The relativistic electrons in the beam could be accelerated by any one of three mechanisms. The first mechanism entails the existence of shock waves in the jet much like the ones formed in the earth's atmosphere when an airplane is flying at a velocity greater than that of sound. Such shocks could result from the collision of the jet with interstellar clouds. They could also form internally if the beam is unsteady or unstable. The knots in the beam would then correspond to regions where the plasma is subjected to strong shocks.

In the second mechanism the energy of turbulence in the beam provides the energy to accelerate the electrons. Instabilities and turbulence can form on the boundary layer between a beam of plasma and the medium that surrounds it. The turbulence could propagate into the beam in the form of waves whose motion would be damped by interaction with the electrons. In interacting with the waves the electrons would be accelerated. The net result of the process would be to transfer the bulk-flow energy of the plasma into the kinetic energy of the waves and then into the acceleration of the relativistic electrons.

The third possibility is that the electrons could be reenergized by collisions with the protons in the beam. The protons in the plasma cannot be detected directly because they do not emit a detectable quantity of radiation. For the same reason, however, they could retain their high kinetic energy much longer than the electrons do. Collisions between protons and electrons would eventually deplete the reservoir of proton energy, but the process would be slow and the reservoir might be large enough to get the plasma all the way to the inner radio lobe.

The process by which the beam of electrons, protons and magnetic fields gets from the inner lobe to the middle lobe is not yet well understood. However it operates, when the plasma reaches the middle lobe, it could have an important role in the formation of the young stars and the emission-line re-

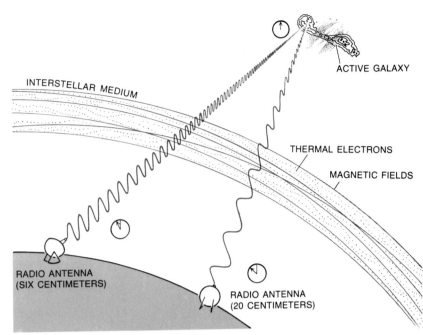

FARADAY EFFECT rotates the plane of polarization of radiation traveling through our galaxy. Radiation from an active galaxy such as Centaurus A is often plane-polarized because it originates in the synchrotron process. As the radiation passes through the interstellar medium in our galaxy it interacts with thermal electrons and magnetic fields. As a result the plane of polarization of the waves is rotated. The degree of rotation is proportional to the wavelength: the longer the wavelength, the greater the rotation. Radio observations of Centaurus A that were made in the early 1960's provided the first direct confirmation of the Faraday effect in the Milky Way galaxy. The observations also helped to strengthen the conclusion that the broad-spectrum radiation emitted by Centaurus A has its origin in a synchrotron process.

gions observed at distances of about 130,000 light-years from the core.

David S. De Young of the Kitt Peak National Observatory proposes the following scenario for the interaction of the radio structures and the optical structures. Eddies in the turbulent layer at the edge of the jet can entrain nearby material, which is mainly gas and dust. As the material is entrained it is heated by shocks in the plasma beam to about 10 million degrees K.

If the elements in the gas surrounding the jet have the same abundances as they do in the sun, which is a reasonable assumption for an evolved galaxy, the gas will cool to 10,000 degrees K. in from 10 million to 100 million years. At a temperature of 10,000 degrees the gas would show optical emission lines like those seen in the northeast quadrant of NGC 5128. Moreover, in the time it took the gas to cool it would have traveled between 33,000 and 330,000 light-years, a range that could put it near the position of the optical emission seen in NGC 5128 and in several other radio galaxies where optical emission lines coincide with radio lobes.

A fraction of the entrained gas would continue to cool and would eventually reach temperatures much lower than 10,000 degrees. The gas would then begin to coalesce by gravitational attraction and stars would form in it. De

Young holds that the clusters of optical emission observed in the northeast quadrant of NGC 5128 support the conclusion that stars are forming in the region of the middle lobe.

The plasma contributes to the formation of stars by entraining ambient matter, but the stars could also help to keep the plasma beam energized as it moves outward from the center of the galaxy. Some of the stars formed in the northeast quadrant would probably have masses greater than 10 times the mass of the sun. Such massive stars have a quite short lifetime: about 10 million years. At the end of their relatively short life the massive stars would explode as supernovas.

The effect of the explosion would be to reenergize the electrons in the beam. Therefore the flow in the northeast quadrant could be self-perpetuating: the plasma beam gives rise to the formation of giant stars, which in turn explode and return their energy to the beam.

This account of how the plasma beam could be energized and how the optical filaments form is based on substantial observational data. Two of the most fundamental questions about Centaurus A and other active galaxies, however, must be answered in large part on the basis of conjecture. The questions are: What is the basis of the galactic activity? How does the central engine work?

It is possible that the activity observed

in NGC 5128 is directly related to the galaxy's unusual structure, which combines an ellipse with a disk. Some 30 years ago Walter Baade and Rudolph Minkowski of the Mount Wilson and Palomar Observatories hypothesized that the form of NGC 5128 is the result of a merger between a spiral galaxy and an elliptical one. Recent theoretical work by Allan D. Tubbs of the National Radio Astronomy Observatory has shown that gas and dust tidally torn from a nearby galaxy can fall into an elliptical galaxy and settle in a shape much like that of the dust lane in NGC 5128.

Such a merging process has been called galactic cannibalism. Cannibalism has been invoked to explain some powerful radio sources other than Centaurus A. Certain clusters of galaxies that include a giant galaxy at their center are five times as likely to be a strong radio source as the average cluster of galaxies. It has been proposed that the giant central galaxy is formed by repeated gravitational encounters deep in the cluster. The encounters generate a frictional force that causes the larger galaxies to spiral slowly inward toward the center of the cluster. Over the lifetime of the cluster the most massive galaxies continue to coalesce, ultimately forming a single massive central unit. The central giant galaxy then cannibalizes the smaller galaxies nearby.

The cause of the radio activity of such giant galaxies is almost certainly the matter accumulated in the process of cannibalism. The accreting gas, dust and stars provide the fuel for the radio engine at the core of the galaxy. A more limited cannibalism in NGC 5128 could also have triggered violent activity in the galactic nucleus and the corresponding radio activity. One problem with the cannibalism hypothesis is that Centaurus A is quite isolated: its only companions within a million light-years are a few dwarf galaxies. Perhaps the galaxy has cannibalized all its neighbors and the nonthermal radio and X-ray emissions are signs of cosmic indigestion in the aftermath of a hearty meal.

How might the central engine utilize the fuel the cannibalism provides? The fuel could probably be consumed most efficiently in the form of a cloud of gas. Observing with the VLA, Jan M. van der Hulst of the Westerbork Observatory in the Netherlands recently found clouds of neutral hydrogen gas within about 500 light-years of the core in Centaurus A. The pattern of the clouds' motion suggests that they are falling toward the nucleus. Such clouds could each year provide the engine with fuel equivalent to at least a tenth of the mass of the sun. If we assume that the engine has an efficiency of 10 percent in converting mass into energy, a tenth of a solar mass per year would be more than enough to

account for the output of the engine, namely the extended radio region.

Fuel for the engine might also come from the atmosphere of stars in the central region of the galaxy. Jack G. Hills of the Los Alamos National Laboratory and others have suggested that stars near the core are subject to large tidal forces, implying that the strength of the gravitational field varies considerably from one side of the star to the other. As a result the star could be broken up or at least stripped of its atmosphere.

Hence matter is available as fuel for the engine at the core of the galaxy. The next problem is to find out how the clouds of gas, which would tend to circle the nucleus, are drawn into it. One way would be for two clouds of gas to collide in the vicinity of the nucleus. In such a collision one cloud might lose angular momentum and fall into the nucleus while the other gained momentum and moved outward.

Having accounted for the fuel and suggested a feeder mechanism whereby the fuel could be delivered to the engine, it is of great interest to consider what is going on in the engine itself. To do so it is useful to have an estimate of the size of the machine. The size of a source of radiation can be estimated by measuring the period over which the intensity of the emission varies. The reasoning underlying such an estimate is as follows. For two widely separated regions of the source to turn on or off simultaneously they must have some means of communication: a physical signal must pass between them. Since the fastest such a signal can propagate is the speed of light, it follows that if a significant fraction of the emission varies in, say, two hours, the source itself cannot be more than two light-hours across.

Although this reasoning is not incontrovertible, it is probably valid for a source with a continuous emitting surface, and we have employed it to estimate the size of the central engine of NGC 5128. Observations of the variability of the X-ray and radio fluxes from the core of Centaurus A show that the emissions vary significantly over a period of months. Faster variations, however, are also observed, with the fastest taking less than 24 hours. Therefore we conclude that the core has components ranging in size from light-months to light-days.

Although any account of what actually goes on in the substructures of the galactic engine is highly speculative, one point seems clear. Whatever the engine's mechanism is, it is not nucleosynthesis, the process of nuclear fusion that supplies the power for most common stars. The assumption of an efficiency of 10 percent for the central engine rules out nucleosynthesis, which has an efficiency of less than 1 percent and would thus

require more fuel and more time to generate the radio structure of Centaurus A than is suggested by recent observations. Moreover, nucleosynthesis yields thermal radiation rather than the nonthermal synchrotron emissions observed in NGC 5128.

A more efficient way for energy to be extracted from matter is for the matter to fall into a strong gravitational field. When that happens, the matter gains considerable kinetic energy. If the energized fuel then collides with structures in the core, energy can be released in the form of high-energy electromagnetic waves and even high-speed particles.

For this to be the mechanism of the central engine of Centaurus A there would have to be a source of a very strong gravitational field at the center of the galaxy. Such a field could be provided by what is referred to as a collapsed object: a black hole with a mass of about a billion solar masses. If the collapsed object exists at the center of the galaxy, it is undoubtedly spinning, since it is difficult to imagine a process that would lead to the creation of a black hole without at the same time giving it considerable angular momentum.

The spinning collapsed object would have three main effects. First, it would provide the well of gravitational potential into which the fuel could fall. Second, its spin axis would orient the entire engine. The spin would cause nearby matter to precess around the engine and form the disk of gas known as an accretion disk. Third, the black hole could have a magnetic field associated with it, and the lines of force in the field could accelerate charged particles to a high energy, extracting energy from the black hole as they did so.

Imagine a cloud of relatively cold gas falling onto the accretion disk. The disk might have the shape of a torus with the black hole in the center. Some of the infalling gas would be collimated within the narrow confines of the black hole and driven back out along the rotation axis of the hole by the radiation pressure of emission from the inner surface of the accretion disk. In this way the disk could both collimate and accelerate the particle beam that forms the inner radio jet.

For the moment this is as far as even the boldest speculation can take us in understanding active galaxies such as Centaurus A. The galaxy remains the focus of intense scientific activity. Because of its proximity and its intrinsic interest, it will undoubtedly continue to be so over the next few years. The current work has amassed much information on the structure and physical processes of Centaurus A. The most intriguing aspect of an active galaxy, however, is the mechanism that underlies its huge emitting region. We believe the study of Centaurus A will help to solve this fundamental problem in astrophysics.

Cosmic Jets

by Roger D. Blandford, Mitchell C. Begelman
and Martin J. Rees
May, 1982

*The violent activity at the center of many galaxies is
manifested in the production of narrow, focused streams
of ionized gas. Some are a few light-years long; others
are a million times longer*

Astronomical observations made with radio telescopes have revealed that the center of many galaxies is a place of violent activity. The most recent discovery about this activity is that it is often manifested in the production of cosmic jets. Each such jet is a narrow stream of plasma (ionized gas) that appears to squirt out of the center of a galaxy, emitting radio waves as it does so. It can be more than a million light-years long. It terminates in an extended blob of radio emission well outside the optical image of the galaxy. The energy content of the blob can exceed 10^{60} ergs, an amount that would be produced by converting entirely into energy the mass of 10 million stars. Although a few jets have long been known from optical observations, it is only in recent years that new techniques in radio astronomy have shown how common jets are in the universe. Plainly jets must be produced under quite diverse conditions, and yet there is no consensus on precisely how they are produced.

Cosmic jets can take many different forms. One of the best-studied examples is the jet associated with the elliptical galaxy NGC 6251, which is 300 million light-years from our own galaxy. In 1977 Peter Waggett, Peter Warner and John Baldwin of the University of Cambridge discovered a long, straight jet emanating from the nucleus of NGC 6251. The jet has an angular width of only three degrees and yet it is more than 400,000 light-years long. Its narrow end coincides with the very center of the galaxy, and there the Cambridge observers found a small, pointlike source of radio emission. In 1978 Anthony C. S. Readhead and Marshall H. Cohen of the California Institute of Technology were able to map this pointlike source. They found that it too consists of a narrow jet with a point source at one end. It was as if a Russian doll had been taken apart only to reveal a smaller version inside. The small jet is colinear with the larger jet, but it is only three light-years long.

A somewhat different-looking jet is known as 3C 449 and is associated with an elliptical galaxy 100 million light-years distant. In 1979 Richard A. Perley and his colleagues, working with the Very Large Array (VLA) of radio telescopes near Socorro, N.M., found that two jets emerge from the center of this galaxy in opposite directions. Both jets show several sharp bends. The northern jet bends quite abruptly to the east about 100,000 light-years from the center; then it bends back and follows a northerly course until it appears to bend again. The southern jet also bends to the east; then it bends back toward the south. There are several jets that show this zigzag behavior.

The phenomenon of jets is not confined to other galaxies. A notable example is a mere 15,000 light-years from the sun in our own galaxy. It is called SS433. From an analysis of optical observations it is inferred that two jets emerge in opposite directions from a binary star system at a speed of 80,000 kilometers per second, or a little more than a fourth the speed of light. The same jets have now been detected at both radio and X-ray wavelengths. Evidently nature manages to produce jets within objects that are as light as a few suns or as heavy as a billion suns (the mass of the nucleus of a galaxy). The jets can be shorter than the distance from the earth to the sun or longer than 10 billion times that distance.

Jets are not new to astronomers. As long ago as 1917 Heber D. Curtis of the Lick Observatory discovered that an optical jet is associated with the large elliptical galaxy M87, which lies in a rich cluster of galaxies in the constellation Virgo. It was not until 1953, however, that R. C. Jennison and M. K. Das Gupta of the Nuffield Radio Astronomy Laboratories at Jodrell Bank near Manchester built the first radio interferometer: two radio telescopes electrically linked so that they can record features in a radio source too small to be distinguished by either telescope alone. Jennison and Das Gupta turned their interferometer toward the radio source Cygnus A, which had earlier been shown by Walter Baade and Rudolph Minkowski of the Mount Wilson and Palomar Observatories to be associated with a distant elliptical galaxy. Much to their surprise they discovered that the radio emission did not emanate from the galaxy itself. Instead it came from a diffuse patch on each side of the galaxy. Radio telescopes of increasing sophistication and power have now been employed to show that the majority of extragalactic sources detected at radio frequencies of less than one gigahertz (a billion cycles per second) have this basic double structure. Since 1953 one of the greatest challenges in extragalactic astronomy has been to uncover the reason. As we shall show, the discovery of radio jets brings the answer much closer.

Throughout the late 1960's and the 1970's a series of increasingly large interferometers were built, particularly at the University of Cambridge, at the Westerbork Observatory in the Netherlands and at the National Radio Astronomy Observatory (NRAO) in West Virginia. The most recent effort is the Very Large Array in New Mexico. It consists of 27 linked radio telescopes that are each 25 meters in diameter, and it can uncover features that subtend an angle in the sky as small as two-tenths of a second of arc. (That angle is subtended by a dime at a distance of 20 kilometers.) In 1971 George Miley and Campbell M. Wade used the NRAO interferometer to reveal the presence of hot spots in Cygnus A. Each such spot is a region emitting intense radiation at radio wavelengths. Typically it lies at the outer extremity of each lobe, or diffuse patch, in the brightest double radio sources (the ones like Cygnus A). Many of the brightest double radio sources show tails or bridges of low-intensity emission that extend backward from the hot spots toward the center of the source, where there is usually a compact focus of radio emission called a core. Most double radio sources are associated (like Cygnus A) with an elliptical galaxy or a quasar (a bright, pointlike object outside our galaxy); in such cases the core is invariably found at the source's optical center.

The resolution of maps made by radio interferometers is limited by the separation of the individual radio telescopes: the greater the separation, the greater the resolution. In the VLA the telescopes are separated by as much as 20 miles. In the technique called very-long-baseline interferometry (or VLBI), which was developed simultaneously with the construction of the modern generation of radio interferometers, the telescopes can be on different continents. The resolution that can then be

achieved is correspondingly greater. Features as small as a thousandth of a second of arc can be distinguished. A thousandth of a second of arc is the angle subtended by a dime at a distance of 4,000 kilometers, by one light-year at the distance of the galaxy NGC 6251 and by 100 light-years at the distance of the farthest quasars.

This great resolution is achieved at a cost. The individual telescopes cannot be linked directly; hence the radio signals from the individual telescopes must be recorded on magnetic tape and compared well after the observations were made. This entails a loss of information, and so the maps that are made with

very-long-baseline interferometry are neither as sensitive nor as detailed (on the scale of the observations) as those made with shorter-baseline interferometry. Still, astronomers have recently been successful at regularly using four or more radio telescopes in very-long-baseline work, and such combinations have gone a long way toward improving the results. The VLBI maps now being made are as good as the maps made with linked telescopes 10 years ago.

Very-long-baseline interferometry has been notably successful in probing the cores of radio sources. It has shown that they usually have features that remain unresolved. These features must

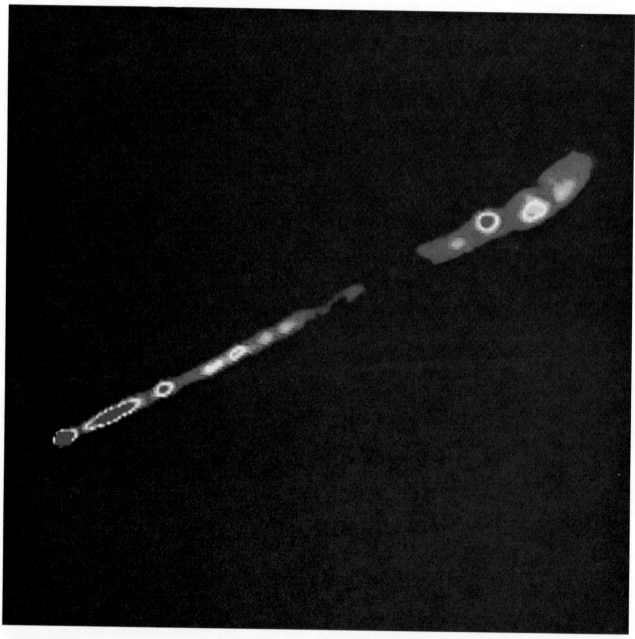

ONE-SIDED JET some 400,000 light-years long is 300 million light-years away from our galaxy. It is the straightest known jet. Its origin (*lower left*) coincides with the center of the elliptical galaxy NGC 6251. Its origin also is known to coincide with a jet only three light-years long that is aligned with the longer jet. The radio emission from NGC 6251 and from the radio galaxies shown on the next page was mapped in arbitrary colors with the aid of the 27 radio telescopes that constitute the Very Large Array (VLA) near Socorro, N.M.

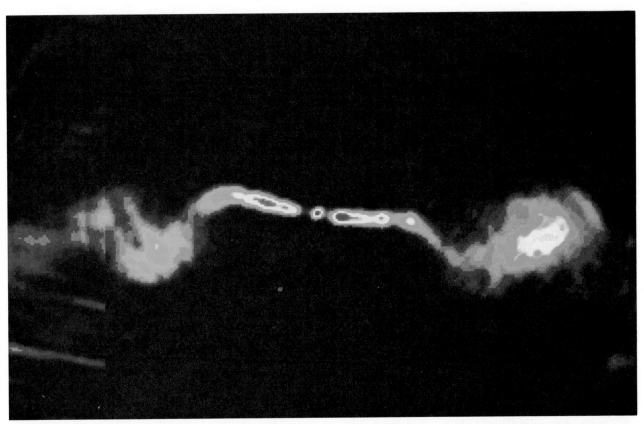

MIRROR-SYMMETRICAL JETS in the radio source 3C 449 emerge in opposite directions from an elliptical galaxy. Each jet bends upward, then downward. Then each emerges into a large lobe of radio emission. It is thought the jets transfer power from the center of the galaxy to the lobes. The mirror-symmetrical bending of the jets may result from the orbiting of the galaxy around a companion.

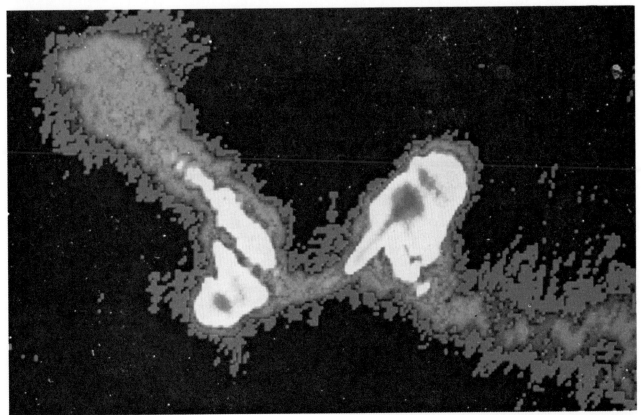

INVERSION-SYMMETRICAL JETS in the radio source 4C 26.03 also emerge in opposite directions from an elliptical galaxy (NGC 326). Here one jet bends upward and the other bends downward. Again, however, the jets each terminate in an extended radio lobe. The inversion-symmetrical bending may result from the precession of the source of the jets, that is, the gyration of the source's spin axis.

therefore be smaller in angular size than a few thousandths of a second of arc. In contrast, the hot spots in the sources generally have no small-scale features. VLBI has also been successful in studying compact radio sources. Such sources are usually associated optically with distant quasars. In each of them the radio flux comes mostly from a small core, not from extended lobes. Nevertheless, radio emission is generally detected at a low level from a region (typically irregular in shape) surrounding the core. The compact radio sources are often found to vary in intensity on time scales ranging from months to years.

Jets have now been found both with linked interferometers and with very-long-baseline ones. They have also been found with a third technique, pioneered at Jodrell Bank, in which radio telescopes are linked by microwave transmissions. More than 70 extragalactic double sources are known to have large-scale jets emerging from the center of the associated galaxy, and the cores of at least six of these double sources show small-scale jets as well. Although a good sample of radio sources has not yet been properly surveyed, some suggestive trends have already emerged.

For one thing, large-scale jets are much more likely to be found in low-power double radio sources than in high-power double sources. Although this is partly due to the fact that the majority of powerful sources are at great distances, so that their jets would be hard to detect, distance cannot explain the trend entirely. For example, no large-scale jets have yet been found in Cygnus A, a powerful double source that is relatively close for such objects. (It is 450 million light-years away.) When a powerful double source does show a large-scale jet, the jet is generally one-sided and no counterjet is found. The jets associated with weaker radio sources are usually two-sided.

A second trend has emerged from the observation that the radio emission from extragalactic radio sources is invariably polarized, that is, the electric fields in the radio waves are directed preferentially along a line on the sky. The reason for this is known. The radio emission is taken to be generated by the synchrotron process, in which electrons moving with speeds close to the speed of light are accelerated by a magnetic field and therefore radiate electromagnetic waves. The electrons are accelerated in a direction perpendicular to the orientation of the field; thus the synchrotron radiation is polarized in that direction.

It follows that by mapping the pattern of polarization for a given radio source one can infer the orientation of the source's magnetic field. Moreover, by measuring the strength of the polarization one can estimate how well-ordered (or untangled) the magnetic field's ge-

ometry is. It turns out that in powerful jets the lines of magnetic field are strung out parallel to the direction of the jet. In weak jets the lines tend to be perpendicular. In jets of intermediate strength one often observes a region of transition from a parallel field in the jet near the center of the associated galaxy to a perpendicular field farther out along the jet.

Even before good examples of radio jets had been found in double radio sources a number of theorists (including us) had argued that the lobes of a double source must be continuously resupplied with energy. And since the radio emission from the core of the double source was clear evidence of continuing activity at the center of the galaxy with which the double source was associated, an energetic link between the center and the extended radio lobes seemed to be implied. The evidence therefore points strongly to the view that jets are streams of gas squirting out of the center of a galaxy. The jets would then supply the lobes not just with energy but with mass, momentum and magnetic flux as well.

After activity begins at the center the jets propagate outward through the galaxy rather like the jet of water that emerges from the nozzle of a hose. The cosmic jets pass first through the interstellar medium and then through the intergalactic medium. The density of the matter there ranges from roughly one hydrogen atom per cubic centimeter to one per million cubic centimeters. Still, the advancing jet must push the matter out of the way; hence the end of the jet moves slower than the gas flowing inside the jet. As a result energy accumulates at the end of the jet; this is the likely interpretation of a hot spot.

The flow of gas in the jet is thought to be supersonic, that is, the speed of the gas is faster than the speed of a sound wave in the gas. As the gas approaches the hot spot, however, it decelerates suddenly. This causes a shock wave to form across the jet. The effect of the shock is important. Before the jet reaches the shock wave most of its energy is in the form of ordered kinetic energy. Passage through the shock converts much of this ordered energy into two forms: the energy of relativistic electrons (electrons moving at speeds near the speed of light) and the energy of a magnetic field. It is entirely natural that the most intense radio emission in a double radio source should be generated where the jet is decelerated by the action of the surrounding gas.

After being stopped at the hot spots, the jet material flows back toward the galaxy. Thus it inflates the large lobes that can be seen in radio maps. The return flow carries with it the relativistic electrons and the lines of magnetic field. A given parcel of the jet spends a relatively short time (roughly 10,000 to a

million years) in a hot spot. Most of the energy of the radio source is therefore contained in the lobes, which probably accumulate gas for 100 million years. The energy content of the lobes is prodigious. In a source such as Cygnus A it must be at least the amount that would be liberated if the mass of 100,000 stars were converted entirely into energy. In still larger sources it can exceed the mass equivalent of a million stars.

Many jets show large bends. In some cases this may be simply a manifestation of an instability that forms in the jet as it bores its way through the intergalactic medium. Anyone who has handled a flexible garden hose will be familiar with the phenomenon. If the hose is laid on the ground in a curve and the water is turned on, the hose will start to writhe. The flow of water in the hose is unstable because the water at the outside of a bend is traveling slightly faster to keep up with the water flowing at the inside of the bend. The pressure on the outside accordingly becomes less than the pressure on the inside, and the configuration of the hose changes in a way that increases its curvature. The physics is basically the same for cosmic jets, although the details are far more complicated if the flow is supersonic.

In a cosmic jet, however, not all bends are produced in a random manner. Consider the sources called radio trails. At low resolution the radio emission from such a source appears not to straddle an optical galaxy as it does in a typical double source but to extend in a long curve on one side of the galaxy. At high resolution the difference is explained. Two jets are seen emerging from the center of the galaxy, but instead of terminating at hot spots the jets are seen to bend steadily so that they both join up with the long, curving trail of radio emission.

What apparently happens in radio trails is that two comparatively low-power jets are formed at the galaxy's center and are then blown sideways by the intergalactic medium as it rushes past the galaxy with speeds known to be several thousand kilometers per second. One is reminded of smoke that emerges from a chimney and then is caught in a high wind. The fact that radio trails are preferentially found in rich clusters of galaxies supports this interpretation of their morphology. (The space between galaxies in the rich clusters is thought to be filled with comparatively dense, hot, ionized gas.) What is most extraordinary about radio trails is that it appears their jets can be bent through almost a right angle by the intergalactic wind without losing their integrity. Thus it is clear that jets are not always unstable. After the jets merge with the trail their gas is presumably brought to a speed matching that of the intergalactic medium. Given, then, the velocity of the medium with respect to the galaxy and knowing the length of the trail, one can measure the age of the radio source. In a typical case the age turns out to be 300 million years.

A related explanation has been advanced for sources such as 3C 449, whose jets show several sharp bends but are basically linear. In general these sources are not found in rich clusters of galaxies. On the other hand, they usually have close companion galaxies. It is likely, therefore, that the radio source corresponds to a galaxy in orbit around its companions. If the velocity of the gas in its jets is not much greater than the orbital velocity, the motion of the galaxy will be "written" on the sky in the form of bends in the jets.

In a sense the bends are an illusion: the gas in the jets does not flow through them. Imagine a photograph of someone watering a garden by swinging a hose. The water hanging in the air has a zigzag pattern, but a given drop of water in the pattern is moving quite ballistically. The shape of the bends in a cosmic jet can be used to infer the three-dimensional positions of the galaxies from their two-dimensional projection on the sky. Still, the bends could be consistent with an orbiting galaxy only if the bends in one of the two jets are related to those in the other by a mirror reflection. If one jet bends to the right on the sky, the other jet must bend to the right at about the same distance from the center of the galaxy. The required mirror symmetry does in fact appear to be present in sources such as 3C 449.

A different class of radio sources exhibits inversion symmetry. In this case a bend to the right in one jet corresponds to a bend to the left in the other. The best example of inversion-symmetrical jets is SS433. Here the source of the jets is thought to be precessing like a top. Again the individual parcels of gas in the jets move in straight lines, but the pattern that should be observed at any one time is that of a corkscrew. Recent radio observations made with the Very Large Array by Robert M. Hjellming of the National Radio Astronomy Observatory and Kenneth J. Johnston of the U.S. Naval Research Laboratory appear to show just that. Perhaps the extragalactic jets that display inversion symmetry come from a source that is precessing just like SS433 but at the center of a galaxy. Such precession would be an important clue to the conditions at the center that give rise to the jets.

We shall now turn our attention to the small-scale jets discovered by very-long-baseline interferometry. Within the past three years several compact radio sources and several of the compact cores associated with extended radio sources have been mapped with sufficient sensitivity to reveal a small-scale jet. In most such cases the jet turns out to have a characteristic morphology. It is a single jet emerging from an unresolved point of intense radio emission. In many instances the single jet points in the approximate direction of one of the source's radio lobes. If an undetected jet points in the opposite direction, it must be at least 10 times fainter than the one that is observed. Readhead, Cohen and their colleagues at Cal Tech have discovered an important correlation bearing on the small-scale jets' alignment. The ones associated with the compact cores of extended radio sources (that is, the sources whose emission comes mostly from double lobes) tend to be well aligned with the lobes. The misalignment is by no more than a few degrees. In contrast, the small-scale jets associated with compact radio sources are noticeably curved, so that they are misaligned with the faint extended structure of the source, typically by angles of 30 degrees or more.

In attempting to interpret this correlation and offer an interpretation of radio sources quite generally, two phenomena are crucial. The first one, called superluminal expansion, has been suspected for some 10 years and has now been rather convincingly demonstrated on recent VLBI maps. Basically superluminal expansion means that features in certain VLBI jets have been observed to move on the sky with respect to the position of the unresolved core from which the jet emerges. The features move at a speed that seems to be in excess of the speed of light. The most convincing case of superluminal expansion has been documented by Timothy J. Pearson and his colleagues at Cal Tech. It involves the quasar 3C 273, which is famous among astronomers for having an optical jet some 60,000 light-years long. The core of 3C 273 emits a VLBI jet, which in accordance with Cohen and Readhead's generalization is bent and misaligned with the optical jet. Over the past four years a bright patch in the VLBI jet has increased its separation from the core by nearly 50 percent. On the hypothesis that 3C 273 is 1.5 billion light-years from our galaxy the bright patch appears to be moving with a speed five times the speed of light.

Astronomers do not contend that the patch is really moving faster than light. That would contravene the special theory of relativity, and there is no need to contemplate such a drastic revision of physics. Instead it is thought we are witnessing an illusion. The bright patch ap-

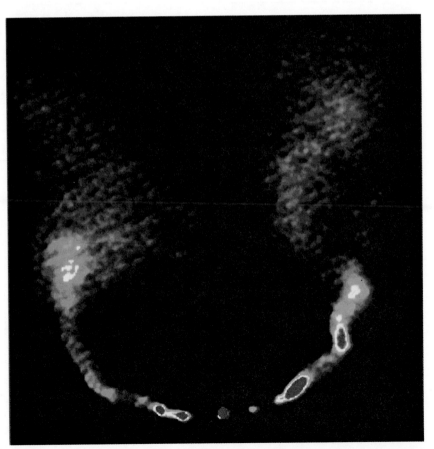

RADIO TRAILS from NGC 1265, an elliptical galaxy in a rich cluster of galaxies, may represent jets blown into a long curve by intergalactic gas that passes the galaxy at a speed of several thousand kilometers per second. The map was made (in arbitrary colors) with the VLA.

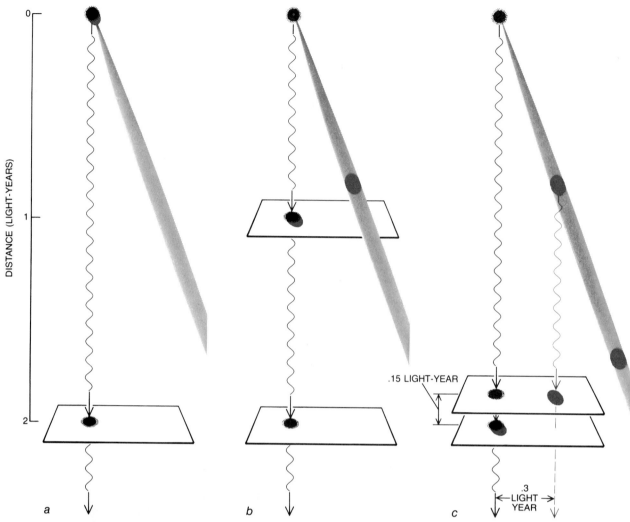

DISTANCE (LIGHT-YEARS)

0

1

2

.15 LIGHT-YEAR

.3
LIGHT
YEAR

a   b   c

SUPERLUMINAL EXPANSION (the apparent motion of components of a radio source at speeds exceeding the speed of light) is thought to be an illusion. Still, it suggests that those components can have speeds near the speed of light. In this example an observer two light-years from a radio source (*a*) is unaware that the source has emitted in its jet a bright knot of ionized gas (*color*). After a year (*b*) the observer still sees a single source; the light showing the emergence of the knot has a light-year to go to reach him. Meanwhile the knot has moved away from the source at an angle of 20 degrees to the line of sight. Its speed is .9 times the speed of light. After two years (*c*) the observer sees the knot emerge. The light the knot emitted in *b* is only .15 light-year behind. Hence in .15 year the observer will see that the knot has moved to .3 light-year from the source. He may mistakenly judge that the knot is moving at twice the speed of light.

pears to be moving faster than light because the light from the patch and the light from the core of the quasar, which serves us as a benchmark, travel along paths of different length and therefore take slightly different amounts of time to reach us. The precise arrangement is still controversial, but the simplest possibility requires that the bright patch be moving at a small angle to the line of sight between the radio source and observers on the earth with a speed slightly less than the speed of light. It will then seem to move superluminally. The important point is that superluminal expansion suggests not speeds faster than the speed of light but speeds not much less than the speed of light for the components of radio sources.

The second phenomenon that must be appreciated before one can analyze

radio sources is relativistic aberration. Suppose a hunter wants to shoot a duck when it is directly above him. He must point his shotgun upward and fire before the duck passes overhead. The shotgun pellets will travel vertically upward to the duck, which will have moved forward to meet them (the hunter hopes). Now consider all of this from the vantage of the duck. It sees the hunter moving toward it, and the pellets, instead of moving vertically, have a horizontal component in their motion. To put it another way, the pellets travel with a slight aberration that inclines their trajectory along the direction the hunter is moving with respect to the duck.

The same thing happens with photons, the particles of electromagnetic radiation. Hence a cloud of plasma that radiates photons equally in all directions will appear to be shining prefer-

entially along its direction of motion. When the plasma is moving relativistically (that is, almost as fast as the photons), the effect is quite pronounced. Take the case of the plasma whose relativistic motion gives the illusion of superluminal expansion in a radio source. Here roughly half of the photons the plasma emits will be radiated within a cone facing into the direction of the plasma's motion with an opening angle of only five to 10 degrees. Furthermore, the photons within the cone will have been made more energetic by a Doppler-shift increase in their wavelength. The net result is striking. If an observer is in the cone, the source will look brighter than a stationary source, typically by a factor between 100 and 1,000. Conversely, if the observer is well outside the cone, the source will be effectively invisible.

We can now advance a powerful unifying interpretation of extragalactic radio sources. Suppose most sources consist of a pair of relativistic jets emerging in opposite directions from the heart of a galactic nucleus. If the jets are more or less aligned with the direction from the source to the earth, we detect the jet that is pointed more or less toward us. The other jet is invisible. The source appears bright and one-sided. The jets end in a pair of extended radio lobes, which radiate photons roughly equally in all directions. They produce the faint halo of extended emission that is seen to surround a compact source. One does not expect the jets to be precisely uniform and straight. Doubtless they exhibit some natural bending and inhomogeneity. In the compact radio sources the bending is exaggerated because our line of sight is almost colinear with the jet. If the inhomogeneities move along the jet at relativistic speeds, they will appear to be moving superluminally.

On this interpretation the extended radio sources are observed at large angles to the direction of their jets. For the most powerful of these sources, the ones such as Cygnus A, the jets are relativistic, so that they are effectively invisible. For sources somewhat less powerful, such as the radio source associated with the galaxy NGC 6251, the jets are mildly relativistic. Nevertheless, the aberration of their radio emission is sufficiently pronounced for only the jet on our side of the center of the radio galaxy to be detected. For weak sources such as 3C 449 the jets are subrelativistic, so that both jets are equally apparent.

The "Doppler beaming" interpretation of extragalactic radio sources accounts for so many diverse observations that it is unlikely to be totally wrong. It must be said, however, that the interpretation is difficult to reconcile with certain observations. For example, the density of the gas in some of the one-sided jets may turn out to be so high that if its speed were relativistic, the power carried by the jet would exceed by a factor of 10,000 the amount of power apparently dissipated by the jet and the source's radio lobes.

In some sources, therefore, it appears that one-sidedness is not the result of relativistic motion. An alternative hypothesis is that the visibility of a jet is related to its stability. In the strongest sources the jets would be stable, invisible conduits that efficiently transport energy from the core to the radio lobes. The jets, however, could be rendered unstable. The instability would readily channel the jets' kinetic energy into the acceleration of relativistic electrons and a greatly enhanced radio emissivity. One-sided jets would arise when only one of the jets is rendered unstable. Two-sided jets would be associated with

low-power sources in which the transport of energy is inefficient. An observational test that will discriminate between the competitive explanations of one-sidedness in radio sources will be to determine in one-sided sources whether

the radio lobe on the side of a one-sided jet is systematically different from the lobe on the side that seems to lack a jet. If this turns out to be the case, the Doppler-beaming explanation will be hard to maintain.

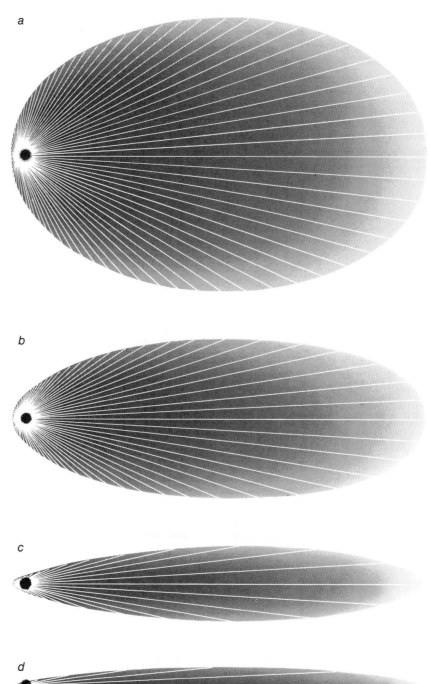

RELATIVISTIC ABERRATION focuses the radiation emitted by an object moving at a speed approaching that of light, so that the object radiates intensely in the direction of its motion. In *a* the emitter (a cloud of ionized gas that intrinsically radiates equally in all directions) is moving toward the right at half the speed of light. In *b* it is moving at .75 the speed of light, in *c* at .94 the speed of light and in *d* at .98 the speed of light. The emitter is rendered invisible except from in front of it. The shape of each pattern shows only the way the intensity of the radiation varies with the angle of emission. Seen from directly in front, however, the emitter in *a* is seven times brighter than a stationary source radiating equally in all directions, the one in *b* is 30 times brighter, the one in *c* is 440 times brighter and the one in *d* is 3,100 times brighter.

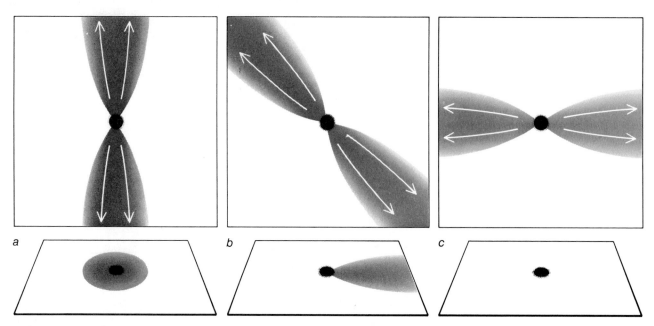

DIFFERENT ORIENTATIONS of cosmic jets may give them different appearances because of relativistic aberration. Here each radio source consists of two jets emerging in opposite directions from the center of a galaxy. The center radiates equally in all directions, but the gas in the jets moves fast enough for the jets to radiate preferentially in the direction of the motion. If the jets are parallel to the line of sight (*a*), a distant observer perceives a core of radio emission surrounded by a radio halo. If the jets are almost parallel to the line of sight (*b*), the observer detects only the core and the nearer jet; the far jet is invisible because little of its radiation is directed toward the observer. If the jets are perpendicular to the line of sight (*c*), the observer detects only the core because aberration makes both jets invisible.

One difficulty plaguing the investigation of cosmic jets is that it is not known how much total power the jets are radiating. We know only how much they radiate at radio wavelengths. The problem is that radio telescopes are far more sensitive than optical telescopes. A few jets have been detected at optical or X-ray wavelengths. This means their optical power must be at least 100 times greater than their radio power. We cannot rule out the possibility that the radio sources are generally most luminous at optical frequencies.

A further difficulty is that it is hard to tell whether a jet delineated by radio emission is identical with the region where ionized gas is flowing. If much of the jet's true cross section is invisible, such quantities as the power and thrust of the jet will tend to be underestimated. The estimates made today present an intriguing puzzle. Some of the most luminous jets appear to become collimated as they move away from the galaxy that emits them. What, then, squeezes the jet? Is it the pressure of the gas that surrounds the jet in the intergalactic medium? Then the heating and collisions in the gas should make the region around the jet a powerful X-ray source. Is it loops of magnetic field that are carried along with the jet? A property of curved magnetic field lines in the jet is that they exert a tension, so that a loop of magnetic field wrapped around a jet can act like a rubber band. One finds it worrisome that the strongest jets, the ones most in need of confinement, have magnetic fields that tend to be parallel to the axis

of the jet rather than perpendicular to it. Still, this problem can be overcome if only the core of the jet is sufficiently luminous to be observable.

To a theoretician the most important question about the jets is what makes them. A rich variety of schemes have been suggested, but none of them commands universal support. One possibility is that jets are made by a nozzle. Imagine a copious supply of hot gas in the middle of a cooler large cloud. The hot gas may result from the presence of exotic astronomical objects such as black holes or pulsars that heat the gas when it falls in their intense gravitational field. Imagine that the cool cloud is spinning. The hot gas will be buoyant and will tend to escape along the directions of least resistance, which lie along the cloud's axis of rotation. The cloud will provide flexible walls for the outflow, and as the gas flows outward its pressure will decrease to match the external pressure. This can lead to an increasingly well-collimated jet.

To be sure, the hypothetical mechanism of a nozzle is in conflict with the observations made of several radio sources. For one thing, it seems likely that the VLBI jets are produced in regions no more than a few light-years in size. Yet if a gas cloud were dense enough to confine the pressure of a jet in such a small place, the cloud would be a spectacular source of X rays. Furthermore, the alignment of the VLBI jet and the large-scale jets in the extended radio sources suggests that the collimating

mechanism must be a good gyroscope: it must have a long "memory" for a fixed direction in space. It is not clear that a gas cloud can be a good gyroscope.

These difficulties are alleviated if the collimation is assumed to happen in the region surrounding a black hole whose mass may lie anywhere between a million solar masses and several billion. The presence of such a black hole at the center of radio galaxies is supported by several independent items of circumstantial evidence. First, some of the most active radio sources show day-to-day changes in their optical flux. This suggests that the flux comes from a region no larger than the distance light can travel in a day. That distance is only 10 times greater than the radius of a black hole whose mass is a billion solar masses. At X-ray energies great changes in flux in less than an hour have been reported. Second, a group of optical astronomers led by Wallace L. W. Sargent and Peter J. Young at Cal Tech have investigated the light from the center of the galaxies M87 and NGC 6251. They have argued that the distribution and the spectrum of the light indicate the presence of a compact mass exceeding a few billion solar masses. Third, a black hole represents the deepest possible gravitational "well." Hence a black hole is best able to effect a high-efficiency conversion of the energy of infalling gas into radiant or kinetic energy.

If the black hole is spinning, it will distort the space around it by an effect that can be predicted by means of the general theory of relativity. The distortion will cause the gas surrounding the hole to precess around the spin axis of the hole. The gas at different distances will precess at different rates. As a result two things will happen. First the viscous drag between neighboring rings of gas will make the gas settle into a disk in the equatorial plane of the hole. Then the same viscous drag that caused the disk to form will make matter drift slowly inward through the disk. The drift will feed jets with mass and will also power the jets by the release of gravitational energy. Presumably the jets will be launched in the directions parallel to the black hole's spin axis. Since the spinning black hole has a great deal of angular momentum, these directions are fairly stable.

The structure of the accretion disk will be governed by the rate at which gas is supplied by viscous drag. If the rate of supply is large, the pressure of the radiation produced in the innermost part of the disk will puff up the innermost part. As a result a thick torus, or donut-shaped volume, of gas will surround the black hole. The surface of the torus near the spin axis of the hole will define two narrow funnels in each of which a jet can be collimated. Each jet would be driven outward through its funnel largely by the pressure of the radiation field inside the funnel and perhaps in part by mechanical energy brought to the surface of the funnel by convection in the gas inside it. One difficulty persists. A disk supported by radiation pressure implies an optical and ultraviolet luminosity far greater than is generally observed at the center of the galaxies associated with double radio sources. There is nonetheless reason to think that in SS433, the jet-emitting object within our galaxy, the gas supply is very large. A torus supported by radiation remains a viable explanation for the production of jets by that source.

A variant on the radiation-supported torus is the gas-supported torus, which can be produced at a low rate of gas supply. Here the radiation generated in the innermost part of the disk escapes freely and cannot be responsible for inflating it. Nevertheless, the viscous friction in the disk is great enough for the gas to liberate gravitational energy by sinking toward the black hole faster than it can radiate energy. This means the gas gets hotter as it falls. The heating causes it to fill a toruslike volume.

The power for the jets need not be derived from the infalling gas but may be drawn from the black hole itself. In particular a significant quantity of mechanical energy can be extracted from a spinning black hole. Suppose the gas flowing toward the hole has a magnetic field. Eventually some of the field lines penetrate the surface of the hole. The lines become loosely attached to the

hole, and as the hole spins it drags the field lines along with it. These lines can therefore do work on the matter near the black hole. Indeed, the field lines attached to the hole are rather like ropes around a windlass. As for the black hole itself, it acts like a giant flywheel that is gradually being spun down.

We have not yet touched on how the cosmic jets are fueled. Fundamentally the most powerful radio sources require at least a few stellar masses of gas each year unless the source is powered in part by a spinning black hole. The normal processes of stellar evolution in the galaxy surrounding the core of the radio

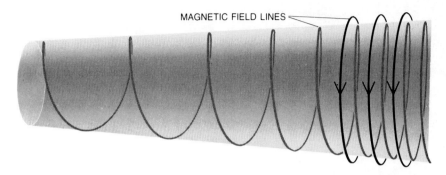

MAGNETIC FIELD LINES

MAGNETIC FIELD LINES in a jet respond to the expansion of the jet by becoming more nearly perpendicular to the direction of flow of the gas in the jet. Basically the field in the jet has a component parallel to the direction of flow and a component perpendicular to the direction of flow. The expansion of the jet spreads the former more than the latter. Thus the loops of a helical field line (representing the superposition of a parallel field and a perpendicular field) can incline toward the perpendicular with distance from the core of the radio source. Perpendicular loops of the field can exert a tension on the jet; they may thus arrest the jet's expansion.

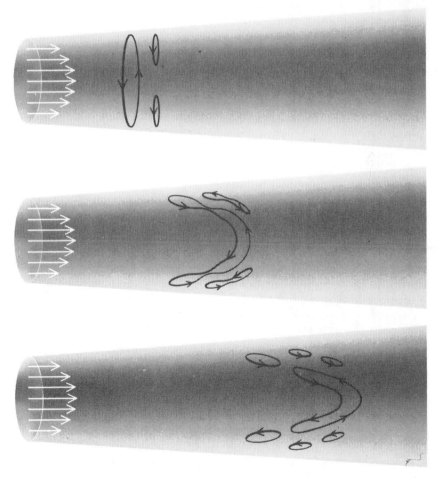

SHEAR IN THE JET has a different effect on the jet's magnetic field: it makes the parallel field dominant. The shear, caused by lack of uniformity in the velocity of the gas in the jet (white arrows), elongates the loops in the perpendicular field. The elongated regions may ultimately break away from their parent loops and close up to become small isolated loops.

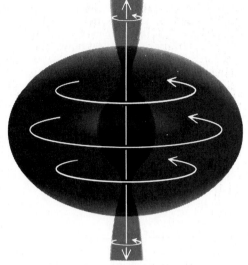

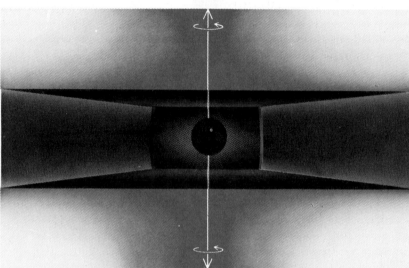

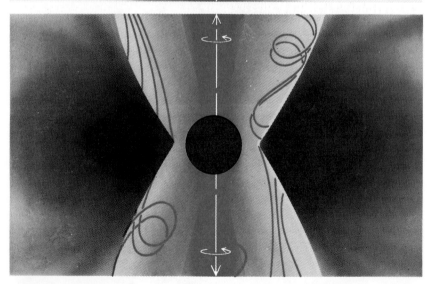

JETS CAN FORM by any one of several hypothetical mechanisms. The nozzle mechanism (*top*) assumes that a rotating cloud of gas surrounds a buoyant cloud of hot gas. The hot gas escapes along the cloud's axis of rotation. The accretion-disk mechanism (*middle*) assumes that the gravitational field of a supermassive black hole collects a gaseous disk. The jets, powered by the pressure resulting from the gas's electromagnetic radiation, are launched along the spin axis of the hole. The accretion-torus mechanism (*bottom*) assumes that radiation pressure (or the pressure of the hot gas) puffs up the inner part of the accretion disk around a black hole. The jets are powered by this pressure, by convective energy in the puffed-up gas or by the spinning black hole, which gives energy to the jets by dragging magnetic field lines through them.

source will meet a certain amount of this requirement. The best available estimates suggest, however, that they fall short by a factor of at least 10. Besides, the violent activity in the core will probably blow gas away. A different suggestion is that the strong gravitational fields at the center of a galaxy will create a dense cluster of stars. Indeed, the stars there may be packed together some 10 million times more densely than the stars near the sun. Such clustering means that stars are much likelier to collide. In addition stars can be torn apart by tidal forces if they come too close to a central supermassive black hole. Still, detailed calculations have shown that it is hard for the required cluster of stars to build up and persist.

A third possibility has observational support in the form of galaxies with double or even multiple centers. It is a process that has been termed cannibalism by Jeremiah Ostriker of Princeton University. Here the center of a "cannibal" galaxy consumes a smaller companion galaxy: a "missionary" galaxy. The outer parts of the missionary are quickly stripped away by tidal forces. The inner parts prove more difficult to digest. Thus the inner parts sink toward the cannibal's center. The idea that cannibalism fuels jets is attractive because radio galaxies are commonest at great distances. They were commoner, therefore, in the past, and specifically at times when galaxies were closer together and likelier to interact. A galaxy was 1,000 times more likely to be a radio source when the universe was a fourth its present age. Moreover, it is known that among nearby galaxies the ones that happen to be in places where the density of galaxies is high are far more likely to be radio sources than the ones that are isolated.

If radio sources indeed are fueled by cannibalism, an explanation of the inversion-symmetrical jets emerges. Suppose both the missionary galaxy and the cannibal galaxy have a central supermassive black hole; then the missionary's black hole will eventually settle into a binary orbit around the cannibal's. Suppose too the missionary's black hole maintains jets that are collimated along the hole's spin axis. Normally the hole would point its spin axis along a fixed direction in space. Now, however, the gravitational field of the companion black hole will distort the space in its neighborhood. As a result a distant observer would find that the spin axis precesses. If the spin axis precesses, so do the jets.

Thanks mainly to the construction of the large linked interferometers and the simultaneous development of very-long-baseline interferometry, radio astronomers have made great strides toward elucidating the structure of the double radio sources, including their cosmic jets. On the interpretive side, however, we remain embarrassingly ignorant. We lack a clear idea of the composition of the jets, of how fast they are moving, of how they are confined and of why they are so stable. Most important of all, we do not really understand how they are made. We have only a wealth of hypotheses. Some are ingenious, some are improbable but many seem quite reasonable.

The pervasiveness of cosmic jets suggests that the raw materials of a deep gravitational well, a supply of gas and some spin are all that are needed for a galactic nucleus to expel its exhaust in opposite directions. Jets are nonetheless heterogeneous. Some appear to be fast, others slow. Some appear to be powerful, others feeble. Some are long, others short. Perhaps theorists have been trying too hard to find one mechanism that accounts for everything. A diverse set of processes might find counterparts in nature and always lead to cosmic jets.

What are the prospects for progress in the study of cosmic jets? On the observational side the prospects are bright. The comparison of extragalactic jets with the jets in our galaxy should yield insight into which aspects of the production and propagation of jets are independent of scale. In addition to SS433 the X-ray source Scorpius X-1 (only 600 light-years away) shows clear evidence of double structure. Moreover, a number of clouds of interstellar gas and dust have offered tantalizing hints that jets may arise in conjunction with the formation of stars.

So far the Very Large Array has mapped only a small fraction of the radio sources accessible to it, and only a very few such sources have been mapped at the highest available sensitivity and resolution. Moreover, the technique of very-long-baseline interferometry has now come of age. The next stage is to construct large arrays of radio telescopes that are dedicated to VLBI. At optical wavelengths one can look forward to the launching of the Space Telescope in 1985. This instrument, which will have more than 20 times the resolving power of comparable telescopes on the ground, will reveal much about quasars and may discover several more jets at optical wavelengths.

On the theoretical side much work remains. Detailed calculations of the gas flow around massive black holes can be done with computers. Such studies are handicapped, however, by our ignorance of the microscopic properties of plasma under cosmic conditions. The experimental approach may prove very fruitful. In particular, aerodynamicists routinely experiment with jets of gas. Typically the jets they make have a velocity no more than a few times the speed of sound in them. Greater speeds may be needed. It is an exciting prospect that the strange and beautiful shapes in the radio sky might be reproduced in a wind tunnel.

Superclusters and Voids in the Distribution of Galaxies

by Stephen A. Gregory and Laird A. Thompson
March, 1982

Red-shift surveys of selected regions of the sky have established the existence of at least three enormous superclusters of galaxies. The surveys also reveal that huge volumes of space are quite empty

Astronomers and cosmologists are much preoccupied these days with explaining the emergence and distribution of aggregates of matter in the universe. How soon after the big bang, the presumed explosion of the primordial atom some 10 to 20 billion years ago, did matter begin to coalesce into the stars and galaxies we see today? Assuming that matter was more or less evenly dispersed before coalescence began, is the universe on the grand scale uniformly populated today by stellar aggregates of one kind or another? Recent observations by several groups of astronomers are helping to answer these questions. Large-scale surveys have verified the existence of superclusters of galaxies: organized structures composed of multiple clusters of galaxies. Each cluster, in turn, may consist of hundreds or thousands of individual galaxies. Although the existence of superclusters has long been conjectured, their confirmation has been accompanied by at least one major surprise: equally large regions of space contain no galaxies at all.

Superclusters are so vast that individual member galaxies moving at random velocities cannot have crossed more than a fraction of a supercluster's diameter in the billions of years since the galaxies came into being. Evidently superclusters offer an insight into evolutionary history that is simply not obtainable with smaller systems. At scales smaller than those of superclusters the original distribution of matter is smeared out by evolutionary "mixing." Astronomers hope that an understanding of the largest structures in the universe will clarify the processes that give rise to structures of all dimensions, ranging downward from galaxies to stars and planets.

It is impossible to determine who first conceived the idea that clusters of galaxies might be members of still larger aggregates, namely superclusters. As one reads old technical papers on extragalactic astronomy one is struck by the similarities between the speculations of 50 years ago and the better-understood concepts of today. What our immediate predecessors lacked were the observational tools that have finally provided the evidence to substantiate some of the early speculations. Although observations in the X-ray, ultraviolet, infrared and radio regions of the electromagnetic spectrum have opened exciting new windows on the universe, it is fair to say that the most important information for cosmology has been collected by telescopes that gather visible and near-visible light.

Even before the invention of the telescope observers could see in the night sky not only planets and stars but also nebulous patches of light. As large telescopes evolved in the 19th century some of the nebulas were resolved into individual stars. Some astronomers contended that such nebulas were huge independent systems of stars outside our own system. In any event the positions of nebulas listed in John Herschel's *General Catalogue* of 1864 and in J. L. E. Dreyer's *New General Catalogue* of 1888 clearly show clustering.

As a consequence astronomers who believed some of the nebulas were external systems began to speculate about their clustering tendency. In 1908 the Swedish astronomer C. V. L. Charlier outlined a cosmology characterized by a hierarchy of clustering. He identified a number of nebular clusters, two of the largest being in the constellations Virgo and Coma Berenices. In 1922 the British astronomer J. H. Reynolds observed that a band of nebulas seemed to stretch from Ursa Major through Coma Berenices to Virgo, a distance of some 40 degrees across the northern sky. Although Reynolds believed these nebulas were within our own stellar system, he may be regarded as the first to identify what is now recognized as the Local Supercluster, of which our galaxy is a member.

By the mid-1920's Edwin P. Hubble of the Mount Wilson Observatory proved that many nebulas are indeed external galaxies. By 1929 he had announced the profound discovery, made in collaboration with Milton L. Humason, that the more distant the galaxy is (as judged by its relative faintness), the more its light is shifted toward the red end of the spectrum. Such red shifts indicate that the galaxies are streaming away from ours (and from one another) as part of an expansion of the cosmos. Today the red shift/distance relation is called Hubble's law. It is the basis of modern observational cosmology.

The value of the red shift, z, is obtained by subtracting the rest wavelength, or non-red-shifted wavelength, of a galaxy's stellar spectral lines from the observed wavelength and dividing the difference by the rest wavelength. The largest red-shift value found by Humason (in the late 1940's) was .2, equivalent to a recession velocity of 60,000 kilometers per second, or 20 percent of the speed of light. There is no general agreement on the value for the Hubble relation, but a plausible one equates a recession velocity of 75 kilometers per second with a distance of one million parsecs, or 3.26 million light-years. Humason's galaxy with a red shift of .2 is therefore about 2.6 billion light-years away. The most remote objects known, the quasars, with red shifts as large as 3.5, are apparently receding at more than 90 percent the speed of light. They are presumed to be about 15 billion light-years away.

In the 1930's both Hubble and Harlow Shapley of the Harvard College Observatory drew attention to the fact that in the northern galactic hemisphere (which includes Virgo and Coma Berenices) bright galaxies are more numerous than they are in the southern hemisphere. Hubble also photographed many galaxies so faint that he believed he had found a probable end to the phenomenon of clustering. "For the first

CORE OF THE CLUSTER OF GALAXIES in the constellation Coma Berenices, a small part of the vast supercluster that includes not only the Coma cluster but also another rich cluster, A1367, is depicted in a color image reconstructed from black-and-white photographs. At least 300 moderately bright elliptical galaxies and galaxies of the type designated S0 can be counted in the original photograph, each a giant collection of tens of billions of stars at a distance of some 300 million light-years. The only prominent object in the picture that is not a galaxy is a bright blue star, slightly above and to the right of the center, that is a nearby member of our own galaxy. The two giant elliptical galaxies near the center are dominated by the ruddy light from much older stars. The picture was prepared at the Kitt Peak National Observatory with the aid of a computer-controlled television monitor in which one can carefully balance colors appropriate to black-and-white images recorded separately in the red, green and blue (or ultraviolet) regions of the electromagnetic spectrum. The original photographs were made by one of the authors (Thompson) with the 1.2-meter Schmidt telescope on Palomar Mountain.

SPIRAL GALAXY NGC 4535 is in the constellation Virgo at a distance of some 60 million light-years. It is a member of the Local Supercluster of galaxies centered on the Virgo cluster. Our own galaxy lies in a sparse cluster called the Local Group, which is now thought to be an outlying member of the Local Supercluster. The red-orange hue of the nucleus of NGC 4535 indicates an abundance of old, cool stars. Blue knots in the spiral arms mark regions where hot young stars have recently formed. NGC 4535 is believed to resemble our galaxy in shape, size and luminosity. The picture was reconstructed from images made with the Palomar Schmidt telescope by Thompson.

time," he said, "the region now observable with existing telescopes may possibly be a fair sample of the universe as a whole."

Another of Hubble's contributions, which is pertinent to our discussion, is his classification scheme for the forms of galaxies. He divided the galaxies into two main classes, ellipticals and spirals, with several subcategories. Elliptical galaxies range from the spherical to the lenticular and are generally devoid of structural features. Spiral galaxies, such as our own, are flattened disks in which spiral arms are usually visible (unless the disk is seen from the edge). Later a third group, the S0 galaxies, were found to have intermediate properties.

By 1950 astronomers could agree on the general characteristics of clusters of galaxies. Several very rich clusters were known, the largest being the cluster in Coma, with more than 1,000 individual members. Clusters had been found to consist predominantly of elliptical and S0 galaxies. No more than half of all galaxies appeared to lie within clusters; the rest, classified as "field" objects, were thought to be isolated galaxies, mostly spirals, that lie well outside clusters. A few astronomers suggested that the Virgo region might consist of more than a single cluster, but Charlier's model of a hierarchy of successively larger clusters seemed doomed by Hubble's counts of faint galaxies.

Gerard de Vaucouleurs of the University of Texas at Austin, who has been studying the brighter galaxies in the northern galactic hemisphere since the early 1950's, was the first to define and describe the Local Supercluster. His work shows that it is centered on the Virgo cluster, about 60 million light-years away, and has perhaps 50 outlying clusters called groups, along with individual galaxies scattered among them. Our own galaxy lies in one of the sparse clusters, in what astronomers call the Local Group, so that it is almost certainly an outrider of the great supercluster.

A second major development in the past 30 years is the growing awareness that the Local Supercluster is not unique. Between 1950 and 1954 the entire northern sky was mapped with the wide-angle 1.2-meter Schmidt telescope on Palomar Mountain. Soon thereafter George O. Abell of the University of California at Los Angeles compiled a catalogue of 2,712 rich clusters of galaxies. Abell pointed out that many of the clusters seemed to be members of superclusters with an average of five or six clusters each. His proposal was disputed, however, on the basis of another catalogue of clusters compiled from the same survey by Fritz Zwicky and his colleagues at the California Institute of Technology. The Zwicky catalogue suggested that the clusters do not form clusters of a higher order. The disagreement

can be resolved by recognizing that the clusters as defined by Zwicky are generally larger than Abell's clusters and include several centers in which galaxies are concentrated.

At about the same time, but on the basis of a different sky survey (one completed at the Lick Observatory), Jerzy Neyman, Elizabeth L. Scott and C. D. Shane of the University of California at Berkeley reported finding huge "clouds of galaxies," their term for superclusters. They also suggested tentatively from their study that every galaxy in the universe belongs to a cluster; there might be no isolated galaxies. More recently a thorough and formal analysis of all available galaxy catalogues has been carried out by P. J. E. Peebles and his colleagues at Princeton University. They have quantified the entire spectrum of galaxy clustering and conclude, among other things, that clusters tend to lie close to one another.

A third major development in the analysis of the clustering phenomenon since 1950 has been the introduction of large-scale red-shift surveys. The first step in such a survey is to measure the red shifts of all galaxies brighter than some faint limit in a selected area of the sky. By applying Hubble's law to the observed red shifts the distance of each galaxy can be estimated. This approach has certain clear advantages over the statistical analysis of existing catalogues of galaxies, which provide only the two angular coordinates of galaxy positions, namely the two that can be measured on the plane of the sky. On the basis of the existing catalogue data the third dimension, distance, can only be inferred approximately from the faintness of the galaxy. In the red-shift-survey method the distance is derived explicitly from Hubble's law. The disadvantage of the method is that whereas thousands of galaxy positions can be derived from a single photograph, spectral red shifts can be obtained only one at a time. The two methods are nonetheless complementary. Catalogue studies can analyze large numbers of galaxies in large volumes of the universe; red-shift studies supply three-dimensional detail in much smaller sampling volumes.

Large-scale red-shift surveys have been made possible by major advances in telescope instrumentation. Although Hubble and Humason had access to the largest telescopes (the 100-inch reflector on Mount Wilson and later the 200-inch on Palomar), the available photographic emulsions were slow compared with current ones. Modern spectrographic cameras usually include an electronic device that intensifies the image by a factor of at least 20 before it enters the detector. For certain observations it is sometimes possible to use digital detectors so sensitive that they are capable of counting individual photons. As a result

of these advances and others astronomers can now record as much spectrographic information in half an hour as Hubble and his contemporaries could record in an entire night.

It is of historical interest that the first red-shift survey was presented at a conference held in 1960 on the application of image-intensifying tubes to astronomy. With one of the new devices Nicholas U. Mayall, working with the 120-inch reflector at the Lick Observatory, had recorded the spectra of 40 of the 82 brightest galaxies lying within four degrees of the center of the cluster of galaxies in Coma Berenices. In 1972 Herbert J. Rood and Thornton L. Page of Wesleyan University completed and extended Mayall's initial survey. Additional red shifts were recorded by Eric C. Kintner of Wesleyan, who then analyzed the enlarged sample in collaboration with Rood, Page and Ivan R. King of the University of California at Berkeley. Their paper represents the first modern, detailed study of the red shifts in a single cluster of galaxies. They reported that the cluster consists predominantly of elliptical and S0 galaxies, some moving with speeds of more than 1,000 kilometers per second, and that they could place no limit on the cluster's size.

Some four years later William G. Tifft of the University of Arizona and one of us (Gregory) extended the Coma Berenices survey to both fainter and wider angular limits. We found that the Coma cluster itself ends three degrees from the center but that a number of galaxies form an armlike projection pointing westward toward the nearby cluster A1367 and perhaps linking up with it. (A1367 stands for cluster No. 1367 in Abell's catalogue. The Coma cluster itself is A1656.) Our analysis emphasized that red-shift surveys yield not only a detailed picture of distant clusters of galaxies but also important information about galaxies that may lie in the foreground. Because the galaxies in the foreground seem to be found in the sparse collections called groups (or clouds if they are even sparser) a single red-shift survey is able to identify collections of many different sizes, from the biggest down to the smallest. Indeed, the sparse foreground samples may have as much to tell about how clusters form as the more dramatic rich clusters. Our analysis also drew attention to the paucity of field galaxies.

In a rapidly moving area of investigation the same observations or similar ones are often made independently by different workers. It happened that Rood and Guido L. Chincarini of the University of Oklahoma, who had been studying galaxies to the west of the Coma cluster, found that the Coma cluster's westward arm was still detectable at a distance of more than 14 degrees. They also suggested that a bridge of galaxies might join the Coma cluster and A1367. At this stage the two of us (Gregory and Thompson) initiated a survey that systematically extended the Coma survey westward all the way to A1367. Our more complete analysis confirmed the existence of a bridgelike connection between the two clusters. Hence a study that had originated with Mayall's observation of 40 galaxies in an area of 16 square degrees ultimately covered 238 galaxies in 260 square degrees and in the process established the existence of a true supercluster.

The Coma cluster lies near the extended pole of our own galaxy, nearly 90 degrees away from the veil of dust and gas that limits visibility in the central plane of the galaxy. In our study of the Coma-A1367 supercluster we decided to secure spectra of all galaxies brighter than magnitude 15, about a million times fainter than Vega, the brightest star in the northern sky. When the galaxies in our sample are simply mapped in two dimensions, as they appear in the sky, one can see two main concentrations: the Coma cluster itself in the northeastern corner and A1367 in the southwestern corner [see illustration on page 93]. Otherwise one has a strong visual impression that the map is made up of many unattached galaxies distributed more or less randomly between the two centers.

If the results of the red-shift survey are now plotted to show how the same galaxies are distributed in the third dimension, that is, according to distance, a quite different picture emerges. For this purpose it is sufficient to use two positional coordinates: radial distance (derived from red shifts) and east-west angles in the sky [see illustration on page 94]. In this picture the galaxies are seen to be much less smoothly distributed.

There are a few small clumps close to our galaxy, represented by the vertex of the wedge. The most impressive feature, however, is a densely populated region stretching all the way across the map at a distance of 315 million light-years. This is the feature that qualifies as a supercluster, since it incorporates the two rich clusters, Coma and A1367, and several less populous clusters that together form a continuous megagalactic structure stretching more than 70 million light-years from end to end.

Somewhat surprisingly the map also shows a few "voids," regions seemingly empty of galaxies. At the time we completed the study we felt confident that the voids were real, but we had doubts about their universality. Conceivably, we thought, they were peculiar to this one region of the sky.

Since superclusters themselves may be as diverse in their structure and content as individual galaxies one would like to have examples other than the Coma-A1367 supercluster before generalizing about supercluster properties. At least three more large systems are currently under study, and all of them exhibit intriguingly different character-istics. The region around the Hercules cluster has recently been investigated by one of us (Thompson) in collaboration with Chincarini, Rood, Tifft and Massimo Tarenghi, working with the two-meter telescopes at the Steward Observatory of the University of Arizona and at the Kitt Peak National Observatory. Once again the observational evidence shows the presence of a complex supercluster occupying a broad band at a distance of between 400 and 600 million light-years. Unlike the Coma-A1367 supercluster, the Hercules system is not dominated by one or two clusters. Nevertheless, it is similar to the Coma-A1367 supercluster in having a vast empty region in the foreground. Perhaps the most surprising aspect of the Hercules system, however, is that its densest clusters are dominated by spiral galaxies rather than elliptical ones. This peculiarity alone makes the Hercules system remarkable.

The third supercluster under study is in the region of the sky that coincides with the constellations Perseus and Pisces. It appears to be a long filament that runs more than 40 degrees across the sky, from the well-known Perseus cluster of galaxies to a small group of galaxies near the elliptical galaxy NGC 383. Our new observations, made in collaboration with Tifft, indicate that the depth of the apparent filament is no greater than its width in the sky. Consequently we can presume not

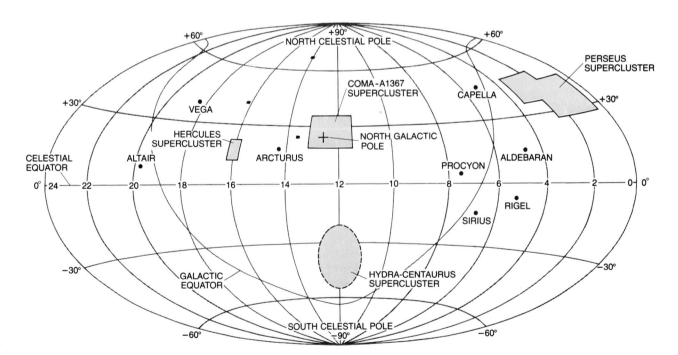

LOCATION OF FOUR SUPERCLUSTERS is shown on a homolographic projection of the celestial sphere. The Hydra-Centaurus cluster in the southern hemisphere has not yet been intensively studied. The first supercluster to be recognized was the Coma-A1367 system. The authors estimate that it holds well over a million galaxies, 280 of whose red shifts have now been measured. It occupies a volume of at least 10^{23} cubic light-years. The three black squares identify regions where a recent survey shows a near absence of galaxies with recession velocities of between 12,000 and 18,000 kilometers per second, corresponding to the interval between 520 and 780 million light-years. If the region between the three sampled sites is equally lacking in galaxies, it would be a void of some 30×10^{24} cubic light-years.

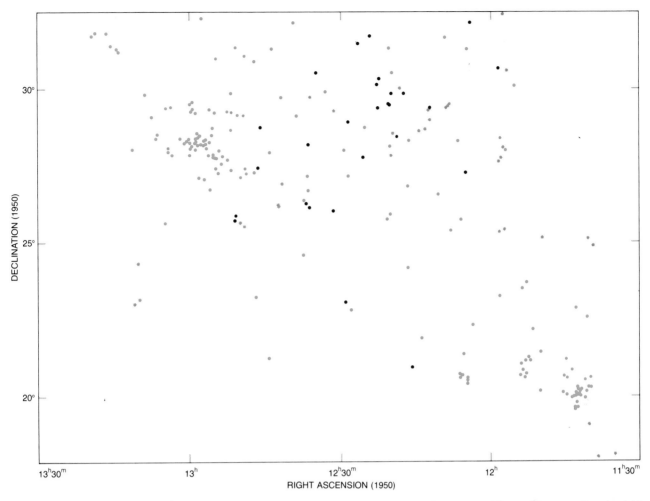

30°

25°

20°

DECLINATION (1950)

13ʰ30ᵐ   13ʰ   12ʰ30ᵐ   12ʰ   11ʰ30ᵐ

RIGHT ASCENSION (1950)

DISTRIBUTION OF GALAXIES in the region of the sky embracing the Coma and A1367 clusters was mapped by the authors in their extensive red-shift survey. The most distant galaxies on the map are shown in color, the nearest are shown in black and those at intermediate distances are shown in gray. The dense concentration of galaxies at a right ascension of 13 hours and a declination of 28 degrees represents the core of the Coma cluster. The smaller aggregation at a right ascension of 11 hours 40 minutes and a declination of 20 degrees represents the A1367 cluster. Because the galaxies elsewhere in the sky seem to be fairly scattered astronomers were misled into thinking most galaxies are sprinkled at random throughout space. A different view of the same region of the sky appears on page 8.

only that the system is a true filament but also that the individual galaxies within the filament have random motions of rather low velocity. There is also evidence that many of the individual galaxies in the Perseus-Pisces system have planes of rotation that are either parallel to the axis of the supercluster filament or perpendicular to it. This observation, if it is confirmed, may tell something about the way galaxies and superclusters are formed.

The three red-shift surveys cover only about 2 percent of the entire sky. Groups at several observatories are trying to obtain a more comprehensive view of the superclustering phenomenon. For example, Jaan Einasto, Mikhel Jôeveer, Enn Saar and S. Tago in the Estonian S.S.R., who independently discovered the Perseus supercluster and the existence of voids, have analyzed the largest existing catalogue of galactic red shifts. Although their catalogue lacks the details of our recent red-shift surveys, it has enabled them to verify on a

larger scale the same features found by the more detailed survey methods.

Similarly, Chincarini and Rood have analyzed the distribution of giant spiral galaxies first observed by Vera C. Rubin, W. Kent Ford, Jr., and their colleagues in the Department of Terrestrial Magnetism of the Carnegie Institution of Washington. The Rubin-Ford survey covers the entire sky but has little detail in any one region. It has nonetheless enabled Chincarini and Rood to verify the existence of the three superclusters we have described and to add a previously unrecognized structure in the southern hemisphere: the Hydra-Centaurus supercluster. The work of Chincarini and Rood and of Einasto, Jôeveer, Saar and Tago strongly suggests that superclusters extend far beyond the regions covered in our red-shift surveys. By their reckoning the Coma-A1367 and Perseus superclusters may occupy an area of the sky 10 times larger than the area we have cautiously proposed.

This hypothesis receives added support from a survey by Robert P. Kirshner of the University of Michigan, Augustus Oemler, Jr., of Yale University, Paul L. Schechter of Kitt Peak and Stephen A. Shectman of the Mount Wilson and Las Campañas observatories. Their survey covers three small regions of the northern galactic hemisphere. In each region they find galaxies with red shifts similar to those observed in the Coma-A1367 supercluster. They also believe they have detected an immense void whose volume may approach 30×10^{24} cubic light-years. In their small search areas, centered near the north galactic pole, there seems to be an almost total absence of galaxies with red shifts of between 12,000 and 18,000 kilometers per second. In some four square degrees of sky, where they would have expected to find about 25 galaxies with red shifts in that range, they found only one such galaxy. The size of the void, which would be by far the largest known, is calculated on the assumption that the region between the sampling

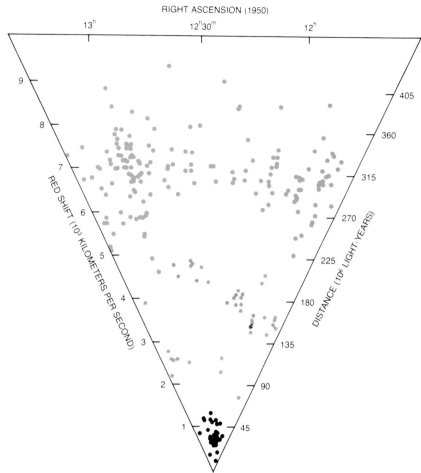

RIGHT ASCENSION (1950)

"PLAN VIEW" OF THE COMA-A1367 SUPERCLUSTER can be drawn on the basis of the authors' red-shift survey. The galaxies are the same ones in the illustration on the opposite page but are distributed according to their velocities of recession as deduced from their red shifts. The Hubble constant, which relates recession velocity to distance, provides the scale at the right side of the triangle. The conversion factor adopted here is that a recession velocity of 1,000 kilometers per second corresponds to a distance of 43.5 million light-years (or 75 kilometers per second per million parsecs). Our own galaxy would be at the lower apex of the triangle at a distance of zero light-years. It can now be seen that the Coma cluster (which is at a right ascension of 13 hours with a recession velocity of about 7,000 kilometers per second and the A1367 cluster (which is at 11 hours 40 minutes and the same recession velocity) are merely the richly populated ends of a continuous band of galaxies that stretches across the sky at a distance from our galaxy of about 315 million light-years. This band of galaxies is the Coma-A1367 supercluster. Note also the huge voids where the survey found no galaxies at all.

points is equally empty of galaxies. The void lies at a distance of between 520 and 780 million light-years.

On the basis of the present surveys we have plotted the three-dimensional distribution of galaxies in the three well-defined superclusters: Coma-A1367, Hercules and Perseus [*see illustration on page 95*]. In this representation our own galaxy is at the center. The three surveys provide wedge-shaped windows looking out into the vastness of the cosmos. The tendency of galaxies to clump is seen to be pervasive. The existence of voids, which we were initially hesitant to credit, can no longer be doubted. The universe might have arranged itself so that the space between clusters was filled not by voids but by many small groups of galaxies. Instead the voids are evi-

dently an integral part of the process of clustering and superclustering.

The study of superclusters is not reserved to optical astronomy; radio and X-ray astronomy are also making fundamental contributions. Radio astronomers are able to detect the presence of intergalactic gas by showing that some radio sources in clusters and superclusters have been distorted by what is presumably a gas of low density but high temperature. If gas fills entire superclusters in the same way as it fills the denser cluster regions, its contribution to the total mass of the supercluster could be tremendous.

X-ray astronomers have already detected extremely hot gas in the direction of distant superclusters. It is unclear, however, whether the emission comes exclusively from the cores of bright

clusters or whether some of it originates in the regions between cluster cores. Jack O. Burns, Jr., of the University of New Mexico and one of us (Gregory) are combining red-shift data from Kitt Peak, radio maps of distorted sources from the Very Large Array radio telescope at Socorro, N.M., and X-ray images from the Einstein X-ray satellite to examine a complete sample of clusters.

Other astronomers are applying the methods of radio astronomy to carry out their own red-shift surveys. The red shifts are determined by observing the displacement in the 21-centimeter radio-emission line of neutral (un-ionized) hydrogen in interstellar space. One survey, conducted by J. Richard Fisher and R. Brent Tully of the University of Hawaii at Manoa, has mapped the galaxies of the Local Supercluster out as far as the nearest large void. The most sensitive radio telescope available for such studies is the 303-meter antenna at Arecibo in Puerto Rico, where work is being done on all three of the superclusters we have described. The astronomers working on these projects include Chincarini, Thomas M. Bania, Riccardo Giovanelli, Martha P. Haynes, Trinh Xuan Thuan and one of us (Thompson). The observations are valuable because they provide not only additional galaxy red shifts but also data on neutral hydrogen at various locations within the superclusters. Although these studies have not advanced far enough to yield new conclusions on conditions within superclusters, they hold great promise for the future.

It has become abundantly clear from the red-shift surveys that the present-day distribution of galaxies is highly inhomogeneous out to a distance of several hundred million light-years. It seems probable that the inhomogeneity extends out to billions of light-years and characterizes the entire universe. We must add the caveat, however, that the universe may contain much matter that is nonluminous. The possible existence and volume of such matter is currently the subject of wide speculation.

If the universe is inhomogeneous today, there is evidence that at very early epochs it was homogeneous. The evidence lies in the fact that the soft background radiation that bathes the earth at microwave-radio wavelengths is remarkably uniform with respect to direction. The prevailing view is that the background radiation represents the expanded and cooled remnant of the hot early universe. Overall the microwave background radiation is smooth to one part in 3,000. Recently, however, some inhomogeneities of small amplitude but extending over large angles of the sky have been detected.

Can the path from homogeneity to the rich assortment of present-day structures be traced? We believe the lack of isolated galaxies and the presence of large voids may provide direct evidence for establishing the relative times of formation of galaxies, clusters of galaxies and clusters of clusters. There are two competing hypotheses. The more conventional model assumes that individual galaxies arose out of a nearly homogeneous primordial soup. The main trouble with this model is explaining how the universe proceeded from its smooth state to the state in which matter was gathered into galaxies. The model assumes that once galaxies formed, small irregularities in their distribution would slowly be amplified by the operation of long-range gravitational forces. The end result of such amplification would be the superclusters seen today.

A competing theoretical explanation was suggested in 1972 by two Russian astronomers, Yakov Zel'dovich and Rashid Sunyaev. In their model the gas of the early universe did not condense into

stars and galaxies immediately. Instead slight but very-large-scale irregularities in the general distribution of the gas grew larger in response to gravitational attraction and became increasingly irregular. Eventually the gas became dense enough to collect into vast sheets of material (dubbed "pancakes"), which then fragmented into galaxies. According to this hypothesis, clusters and superclusters form first as concentrations of gas, and only then do galaxies appear.

Do either of these models find support in the observations we have made of superclusters? Since the Zel'dovich-Sunyaev model requires all galaxies to have formed in clusters or superclusters, field galaxies, or random stragglers, should be rare. If the conventional model is correct and galaxies can arise almost anywhere at random, only later to be shepherded by gravity into groups or clusters, stragglers should be rather common. Actually the only populations

of isolated galaxies we have discovered in our red-shift surveys are galaxies scattered within the boundaries of superclusters. Moreover, the voids are genuinely empty.

We suggest that the isolated galaxies scattered within superclusters were once members of small groups that were subsequently disrupted by collisions within the dense superclusters. It seems realistic to suppose at one time all galaxies were members of groups or clusters. In sum, the observed distribution of galaxies within superclusters and the existence of huge voids between superclusters are entirely consistent with the Zel'dovich-Sunyaev model. Advocates of the alternative model are searching for support in computer simulations they hope will show how small-scale irregularities can grow into large ones by random processes. How the debate will turn out is not clear.

In describing the filamentlike Perseus-

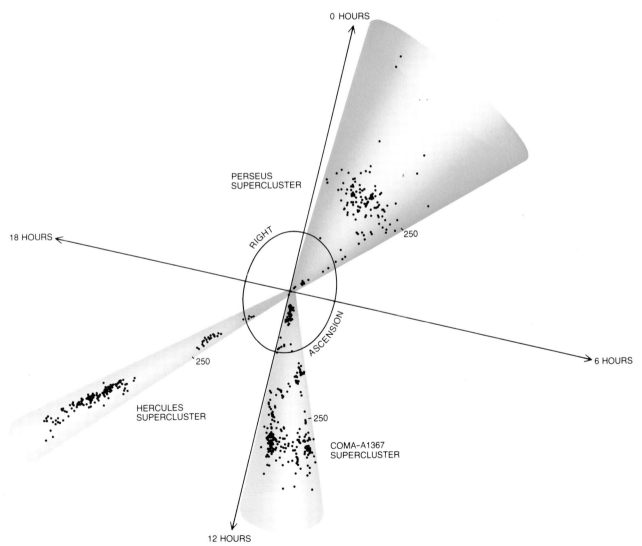

THREE DIRECTIONS IN THE SKY have now been intensively studied in galaxy red-shift surveys. In this projection the solar system is at the apex of three cones that embrace the galaxies that have been surveyed in each of the three regions where superclusters of galaxies have now been identified. The scale marker adjacent to each cone gives the distance from the solar system in millions of light-years. The scale allows for foreshortening in the projection. Thus far only about 2 percent of the entire sky has been surveyed in this detail.

Pisces supercluster we suggested the possibility that the rotation axes of some galaxies were correlated not only with the rotation axes of other galaxies but also perhaps with the gross structure of the supercluster filament. The idea has received support from a recent study by Mark T. Adams, Stephen E. Strom and Karen M. Strom of Kitt Peak, who find similar rotational correlations in the combined data from several flattened clusters. If such correlations are confirmed, supporters of the conventional model of galaxy formation would probably face insurmountable obstacles in producing an explanation. The random statistical processes in the conventional model are not conducive to generating organized rotational motion over any large scale. The Zel'dovich-Sunyaev model, on the other hand, would readily explain such correlations.

What are the prospects of resolving such issues in the near future? One of the most promising avenues of inquiry is continuing improvement in the measurement of the primordial microwave background radiation. The very slight irregularities observed in this radiation are evidence for the existence of structure in the universe from the earliest epochs. The upper limits for the observed smoothness are close to the limits needed for a test of the two competing models of galaxy formation. Perhaps the next 10 years will show that neither model is satisfactory, in which case astrophysicists will have to rethink matters altogether.

Our final comments concern the very concept of a supercluster of galaxies. First, is supercluster the appropriate term? To many of our colleagues the term should be limited to aggregates that are bound together by gravity. It is not clear from the observations that this condition is satisfied. Our own view is that the term supercluster properly describes a present-day aggregate of galaxies that is well separated from any similar structure. No dynamical relations among supercluster members are necessarily implied.

A second comment concerns the universality of superclusters. We have now established that every nearby richly populated cluster in the Abell catalogue is part of a supercluster. We speculate, therefore, that a necessary condition for the formation of a rich cluster is the presence of companion clusters. Finally, we want to leave the reader with a proper sense of the dimensions of superclusters. The Coma-A1367 supercluster is more than 300 million light-years from our galaxy, yet even when it is viewed from such a tremendous distance, it still stretches at least 20 degrees across the sky, through the constellations Coma Berenices and Leo. Chincarini and Rood argue, moreover, that the same supercluster could be as much as 10 times bigger. For astronomers and cosmologists organized structures of such vastness leave plenty of room for study.

Quasars as Probes of the Distant and Early Universe

by Patrick S. Osmer
February, 1982

The light from most of these enigmatic objects was emitted 15 billion years ago. Therefore they are a unique clue to how the universe looked when it was only a fourth its present age

Nineteen years after their discovery by Maarten Schmidt quasars are still one of the great enigmas of astronomy. Although their nature remains controversial, their description is not: quasars are starlike objects with a large red shift. The light emitted by them is strongly displaced toward the red end of the spectrum, signifying they are receding at a substantial fraction of the velocity of light. If quasars are as remote as their large red shifts would indicate, they must be much farther away than ordinary galaxies, whose own red shifts betoken the general expansion of the universe. A quasar can be 1,000 times brighter than an entire galaxy of 100 billion stars. The light from the most distant quasars started on its journey when the universe was only a fourth its present age and has taken 15 billion years to reach us.

It is not my intention in this article to discuss the difficult physics of the nature of quasars. Rather I want to concentrate on what quasars can tell us about the distant universe and its condition at an early epoch. I shall adopt the working hypothesis that quasars are the extremely bright nuclei of galaxies otherwise too distant to be observed. Around some faint, relatively nearby quasars one can detect evidence of a galaxy, but the luminous, distant quasars can be distinguished from stars only by their spectral properties. Distinguishing quasars from stars has presented astronomers with a major challenge. A single wide-field photograph of the sky made with a large Schmidt telescope (named for the optical system designed in the 1920's by Bernhard Schmidt) will show at least 200,000 stellar images, of which only a few hundred will be quasars.

The original quasars were found by optical astronomers who were trying to identify celestial radio sources. As radio techniques improved, the positions assigned to the radio sources became increasingly accurate and in many instances pointed to particular stellar images on photographic plates. Because normal stars could not be detected as sources of radio emission with the equipment then available the coincidence of a radio source with a stellar object was a good way to separate quasars from stars. As a result the majority of the quasars first identified were strong radio sources, hence quasi-stellar radio source, or quasar.

It was soon discovered, however, that a much larger population of quasars were weak radio emitters and therefore had escaped detection by radio telescopes. Allan R. Sandage, working with the 200-inch Hale telescope on Palomar Mountain, found that quasars emitted much more ultraviolet radiation than ordinary stars and therefore could be identified by comparing the stellar images on ultraviolet-sensitive photographic plates with the corresponding images on ordinary plates, which respond chiefly to light at the blue end of the visible spectrum. Sandage showed that quasars that are quiet at radio wavelengths greatly outnumber those with strong radio emission.

Recently a new optical method has been developed for discovering quasars of large red shift that is complementary to the method based on their ultraviolet brightness. The method developed fortuitously from a project Malcolm G. Smith and I began several years ago when we undertook a search for certain galaxies whose spectra exhibit strong emission lines rather than absorption lines, which are more characteristic. The search was conducted with the 60-centimeter Curtis Schmidt telescope at the Cerro Tololo Inter-American Observatory in Chile, in conjunction with two larger reflecting telescopes of 1.5 and four meters. Smith planned to photograph the spectra of faint objects over a wide field by placing a newly acquired thin prism over the entrance aperture of the Schmidt telescope.

The method of spreading the normally pointlike images of stars into spectra by means of an objective prism has had a long and distinguished history in stellar astronomy. What was new about Smith's method was the combination of a prism of low dispersion (to conserve the light from faint objects) and newly available photographic plates with fine grain and high contrast. This combination, together with the exceptional observing conditions at Cerro Tololo, made it possible to record spectra of fainter objects than had been achieved with a telescope of the Curtis Schmidt's modest size. Once emission-line galaxies had been identified by this method their spectral features could be examined in detail in the high-dispersion spectra yielded by the larger telescopes.

On searching the wide-field Schmidt plates Smith noted that in addition to the emission-line galaxies we had hoped to find there were a few objects, possibly quasars, with emission lines at unexpected positions in their spectra. When we examined these objects more closely in the larger telescopes with a Vidicon (intensified television camera) spectrometer I had been adapting for the project, we learned that they were indeed quasars of large red shift. Smith's technique thus provided a direct and efficient method of distinguishing quasars from stars in wide-field plates. In addition the method was more efficient for distinguishing quasars with a large red shift than the one based on ultraviolet brightness.

Smith subsequently surveyed a long strip of the southern sky with the Curtis Schmidt telescope. In addition he and Arthur A. Hoag of the Kitt Peak National Observatory extended the survey to fainter magnitudes in small fields of the Schmidt survey by putting a transmission diffraction grating (equivalent to a prism) at the prime focus of the Cerro Tololo four-meter telescope. I continued to examine the quasars uncovered in the two surveys with the

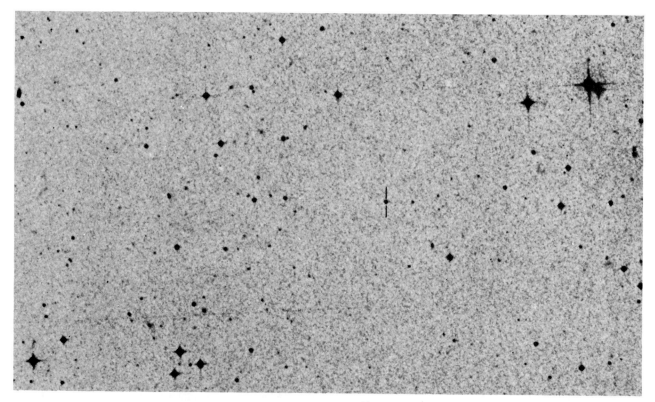

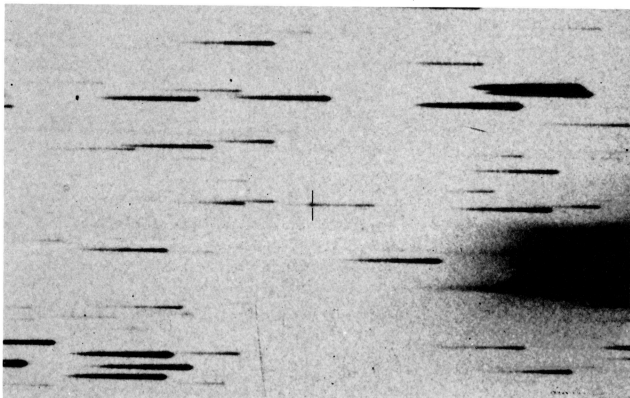

METHOD FOR FINDING QUASARS is illustrated by these two photographs made with the 60-centimeter Curtis Schmidt telescope at the Cerro Tololo Inter-American Observatory in Chile. In a single direct photograph of the sky quasars and stars are indistinguishable because both appear as points of light, as in the negative print at the top of the page. When a thin prism is placed over the aperture of the Schmidt telescope, the image of each object is dispersed into a spectrum, as in the negative print of the same part of the sky at the bottom of the page. In the bottom photograph the longer wavelengths at the red end of the spectrum are at the right and the shorter wavelengths at the violet end of the spectrum are at the left. The spectra of the stars are either devoid of features or exhibit white bands indicating that some of the stellar emission has been absorbed. The quasar, indicated by bars in both photographs, is conspicuous because of a strong emission line, the Lyman-alpha line of hydrogen, that has been shifted from its normal position at 1,216 angstrom units in the far-ultraviolet part of the spectrum to the threshold of the visible spectrum at 3,720 angstroms, corresponding to a red shift of 2.06. This particular quasar, QO 149–397, was one of the earliest discovered by Malcolm G. Smith with the prism on the Curtis Schmidt.

more sensitive Vidicon spectrometer. The combined observations are the basis for the new results I shall describe below.

The spectra of quasars are quite unlike the spectra of all other astronomical objects. The large red shift brings into view regions of the far-ultraviolet spectrum never before recorded by ground-based telescopes. The redshift value, often designated Z, is obtained by subtracting the rest wavelength (the non-red-shifted wavelength) of an emission line from its observed wavelength and dividing the difference by the rest wavelength. The strongest feature in quasar spectra is the Lyman-alpha line of atomic hydrogen, shifted by a factor of as much as 4.5 from its normal position at a wavelength of 1,216 angstrom units in the ultraviolet to about 5,500 angstroms, the wavelength of greenish-yellow light. In such a case Z would be 3.5 (5,500 minus 1,216 divided by 1,216). Emission lines of oxygen, nitrogen and carbon with normal ultraviolet wavelengths between 1,034 and 1,549 angstroms are also prominent in the spectra of quasars with large red shifts.

Moreover, in many instances the emission lines are wide, an indication that some of the gas surrounding the quasar is moving at velocities as high as 10,000 kilometers per second. The physical conditions deduced from the intensities of the various lines show that the gas is hotter than the gas in normal nebulas and that the central source in the quasar does not radiate at all like a normal star. For this discussion, however, the main point is that the Lyman-alpha line is the strongest feature in quasar spectra and therefore is the one most easily detected on the Schmidt plates made through an objective prism. For this reason the objective-prism method favors the detection of quasars with large red shifts.

If red shifts are interpreted as velocities of recession, the red shift needed to displace Lyman-alpha radiation to 3,648 angstroms in the near-ultraviolet region of the spectrum, equivalent to a red shift of 2, corresponds to 80 percent of the velocity of light. OQ 172, the quasar with the largest red shift known, 3.53, is evidently receding at 91 percent of the speed of light. On the same scale stars within galaxies have velocities of .1 percent of the speed of light, and nearby galaxies are moving away from us at no more than 1 percent of the speed of light.

In more than a quarter of a century of measuring galactic red shifts with the 100-inch telescope on Mount Wilson and the 200-inch telescope on Palomar Mountain the largest value recorded by Milton L. Humason was a red shift of .2, obtained in 1949. Another 11

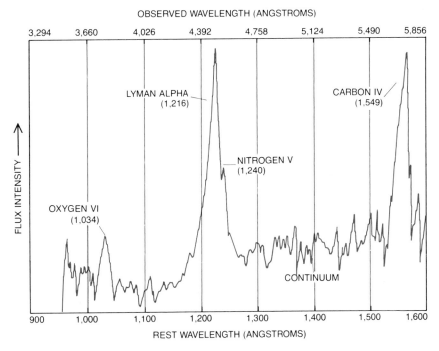

SPECTRA OF QUASARS disclose that the emitting atoms are more highly ionized (stripped of more electrons) than the atoms in nebulas around hot, young stars in our own galaxy. The emission lines are also strongly shifted toward the red end of the spectrum from their rest, or un-red-shifted, wavelength. This spectrum of the quasar QO 453–423 was made with the Vidicon camera system on the 1.5-meter telescope at Cerro Tololo. The strong Lyman-alpha line of ionized hydrogen has been red-shifted to 4,451 angstroms, in the blue part of the visible spectrum, from its rest wavelength of 1,216 angstroms in the far ultraviolet. This corresponds to a red shift of 2.66 (obtained by subtracting 1,216 from 4,451 and dividing by 1,216). The spectra show that emissions from oxygen VI, nitrogen V and carbon IV are comparably redshifted. Roman numerals are one higher than the number of electrons stripped from the atoms.

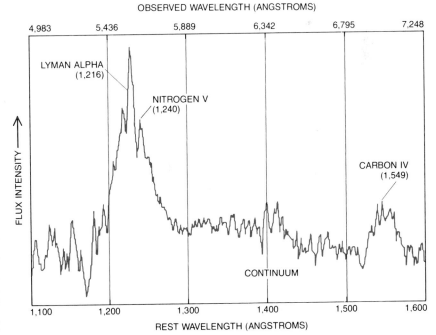

LARGEST RED SHIFT of any known celestial object, 3.53, is that of quasar OQ 172. Its spectrum, made with the four-meter telescope at Cerro Tololo, shows the Lyman-alpha line to be shifted from a wavelength of 1,216 angstroms to a wavelength of 5,508 angstroms in the green part of the visible spectrum. The numerous absorption lines to the left of the Lyman-alpha line are caused by gas clouds lying in the line of sight between the quasar and the solar system.

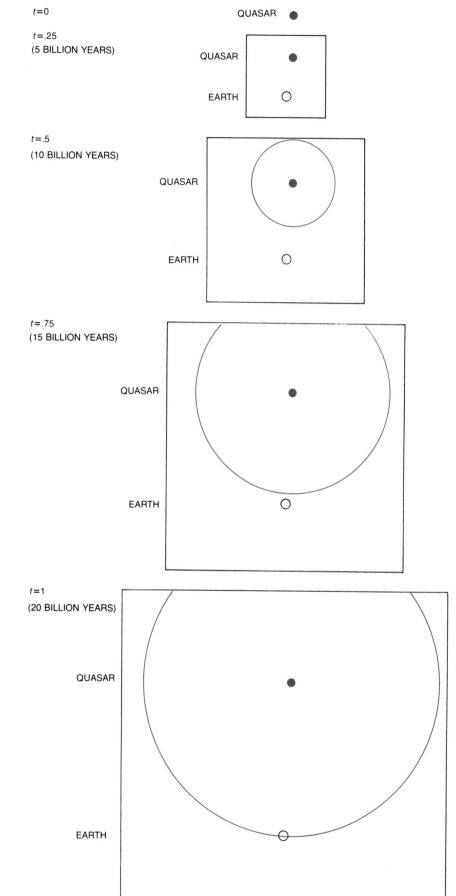

t=0

t=.25
(5 BILLION YEARS)

QUASAR

QUASAR

EARTH

t=.5
(10 BILLION YEARS)

QUASAR

EARTH

t=.75
(15 BILLION YEARS)

QUASAR

EARTH

t=1
(20 BILLION YEARS)

QUASAR

EARTH

years elapsed before Rudolph Minkowski, also working with the 200-inch telescope, pushed the limit out to .46, a record that stood for many years. Within two years of Maarten Schmidt's discovery in 1963 of the first quasar, 3C 273, with a rather modest red shift of .158, the red-shift "barrier" of 2 was broken. The current maximum value of 3.53 was recorded in 1973.

If the quasar red shifts are interpreted as being due to the expansion of the universe, they are not nearby objects that happen to be moving at high velocity but are extremely distant objects. This is a direct consequence of Edwin P. Hubble's conclusion in 1929, based largely on Humason's observations, that galaxies are receding from us (and from one another) at velocities proportional to their distances. On the basis of the current scale of the universe a galaxy with a red shift of only .01 (or an apparent recessional velocity of 3,000 kilometers per second) will be 200 million light-years away. This is already a considerable distance; the nearest spiral galaxy, the Andromeda galaxy, is only about two million light-years away. Quasars are vastly more distant. The quasars with the largest red shifts approach the horizon, or limit, of the observable universe.

Measuring distances in astronomy has always been a problem. Only the nearest stars have truly been measured by trigonometry, as a surveyor measures distances on the earth; everything beyond those stars has been extrapolated. No one contends that red shifts reveal the absolute distances of the galaxies, but it is generally accepted that the red shifts are a good measure of relative distances.

The concept of an expanding universe implies that an object with a large red shift is observed as it existed long ago. The light now reaching us from a quasar with a red shift of 3 was emitted some 15 billion years ago. The light from a quasar with a red shift of only 1 has been traveling for 10 billion years, or half the age of the universe.

A further consequence of an expanding universe is that the red shift supplies a direct measure of the expansion fac-

SCHEMATIC PICTURE of the universe at different epochs illustrates how quasars can supply clues to conditions existing in early epochs. The light now reaching the earth from a quasar with a red shift of 3 was emitted some 15 billion years ago, only five billion years after the hypothetical big bang that initiated the universe's expansion from a dense state (*t* = 0). At an age of five billion years the universe was only a fourth its present size. The colored circles show the distance covered by the quasar's radiation as the universe expands and the quasar and our galaxy move apart. The radiation finally reaches telescopes on the earth.

tor. When the light was emitted from a quasar with a red shift of 3, the universe was only a fourth as large as it is now, that is, everything was four times closer together. The distribution of quasar red shifts therefore supplies many clues to the structure and character of the early universe. I hasten to add that the details of the mathematical relations between the observed red shifts and the properties of the universe are somewhat complicated and certainly controversial, given the sweeping assumptions that are drawn from still limited observational data. Nevertheless, certain aspects of the inquiry are relatively independent of the fine points, and we shall concentrate our attention on them.

In the early years of quasar investigations many astronomers hoped the quasars would help to decide which of the many possible hypotheses about the evolution of the universe was correct. That hope was not fulfilled, but in the late 1960's Maarten Schmidt discovered a remarkable property of distant quasars: they are far more numerous in space at large distances than they are in our vicinity. At a red shift of 2, or about 13 billion years ago, their density was 1,000 times greater than it is now. Evidently whatever process gave rise to quasars must have been extremely active when the universe was young and has subsided to practically nothing today.

At the time of Maarten Schmidt's work on the spatial density of quasars the largest red shift known was 2.88. Enormous as this value was compared with what had been expected only a few years earlier, his result suggested that quasars with much greater red shifts should be plentiful. Why were quasars with red shifts of 3 not being detected? Both Schmidt and Sandage called attention to this point and suggested that quasar red shifts might have a limit.

The implications of such a limit were extremely important. The limit implied that for at least one type of object astronomers were seeing to the edge of the universe. It looked as if quasars had formed suddenly in a great burst of activity about 15 billion years ago. Such a concept, of course, would greatly influence ideas about the fundamental nature of quasars. If it is also assumed that quasars represent enormously energetic processes in the center of galaxies, the red-shift limit has important implications for the evolution of galaxies themselves.

Naturally astronomers kept searching for quasars of ever greater red shift. An obvious concern was that the apparent limit might be the result of some selection effect in the methods of identification, even though the known quasars had been found by quite heterogeneous surveys based on diverse methods. The major point of contention was whether or not quasars with red shifts larger than

2.5 or 3 would continue to be brighter in the ultraviolet than ordinary stars are. The quasar with a red shift of 2.88 had no particular ultraviolet excess, and since its strongest emission line, the Lyman-alpha, is shifted into the middle of the visible spectrum, it and other objects like it would be hard to distinguish from normal stars.

Actually when the quasar with a red shift of 3.53 was found in 1973, it proved to be red in color. It was identified only because radio astronomers had determined its position with high accuracy. If it had been a radio-quiet quasar, it would not have attracted attention. Its discovery, combined with doubts about the completeness of earlier searches, kept the reality of a red-shift limit very much an open question.

The development of our objective-prism method at Cerro Tololo offered a new approach to the problem. Since the method was based solely on the detection of emission lines, it was completely independent of quasar color, ultraviolet or otherwise. Although

the method inevitably had biases of its own, at least they were different from those of all other searches.

The objective-prism method was capable of detecting the Lyman-alpha line in quasars as far into the red end of the spectrum as photographic emulsions allow. The emulsion selected by Hoag and Smith for their deep survey with the four-meter telescope would respond to the Lyman-alpha line as far into the red as 6,900 angstroms, equivalent to a red shift of 4.7. Nevertheless, none of the 71 new quasars found in their survey was found to have a red shift greater than 3.45, which was close to, but not surpassing, the previous limit. Suggestive as this result was, Robert F. Carswell of the University of Cambridge and Smith subsequently showed that the Hoag-Smith survey did not have enough sensitivity for red shifts greater than 3.5 to require that their density actually decrease. Therefore a new survey was in order.

I made observations with the Cerro Tololo four-meter telescope in a program optimized to find quasars with red

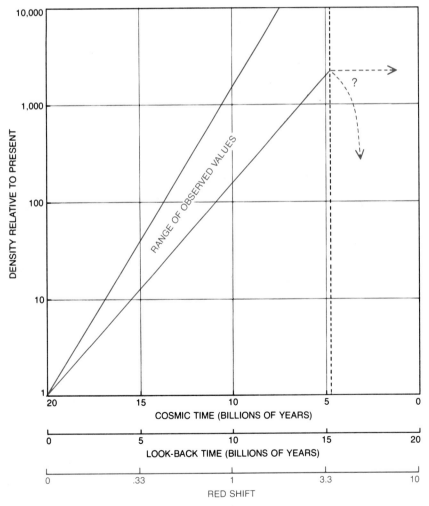

DENSITY OF QUASARS 15 BILLION YEARS AGO was more than 1,000 times greater than it is today, according to calculations made by Maarten Schmidt, who discovered the first quasar in 1963. The absence of quasars with a red shift larger than 3.53 implies astronomers have probed to the earliest epoch of quasar formation. Some 1,500 quasars are now known.

CENTAURUS CLUSTER OF GALAXIES, one of many such clusters, is shown in this photograph made with the four-meter telescope at Cerro Tololo. The Centaurus cluster, which is about 225 million light-years away, harbors some 250 major galaxies, typically separated by about 700,000 light-years. An understanding of the clustering process should help to show how galaxies formed in the early universe.

shifts between 3.7 and 4.7, if they existed. Working with a transmission grating of greater sensitivity in the 6,000-angstrom region in conjunction with a filter to eliminate the light of the night sky at wavelengths of less than 5,700 angstroms, I was able to detect quasars three times fainter than Hoag and Smith could in the crucial 5,700-to-6,900-angstrom region. I found 15 objects with emission lines, any one of which could have been beyond the 3.5 limit. Subsequent observations with the Vidicon spectrometer, however, showed that in no instance was the Lyman-alpha line the one detected on the photographic plate. In five objects the line turned out to be from carbon in quasars with red shifts between 2.8 and 3.4; all the remaining objects were either quasars with red shifts near 1 or galaxies with red shifts of about .2. Although these results were disappointing in one sense, they gave confidence that quasars with a red shift of 4 would have been seen if they were there. All the new quasars were discovered by the detection of lines that were weaker than Lyman-alpha, a complete reversal of the previous surveys with the objective-prism method.

My subsequent quantitative analysis showed that the new survey results could indeed be understood only if the space density of quasars beyond a red shift of 3.5 was a factor of three or more below the value seen at a red shift of 3. The limit does not preclude that a few quasars may eventually be found at greater distances, but it does show convincingly that we have seen the turnover point.

Supporting evidence in favor of the red-shift limit is provided by the failure of radio astronomers to find more distant quasars, even though they now have highly accurate positions that allow them to ignore the color of quasar candidates. Recently X-ray astronomers have entered the field with data collected by the X-ray satellite, the Einstein Observatory. Although the satellite's X-ray telescope readily imaged OQ 172, the most distant quasar, and several other quasars with red shifts of about 3.1, it found nothing to break the record. The next big advance may come with the Space Telescope, which is scheduled to go into earth orbit in 1985. It may tell us just how steep the cutoff is near the limit.

Our work at Cerro Tololo indicates that the density of quasars in space continues to be large up to red shifts of 3.2. The limit at 3.5 implies an abrupt change in the properties of the universe. One of the simplest explanations is that quasars suddenly formed about 15 billion years ago, which would certainly have been a remarkable occurrence in the evolution of the universe. Alternatively an absorbing screen of dust or some other kind of material may be present at a red shift of 3.5 and block our view of the more distant quasars. For the universe to be transparent on one side of the limit and opaque on the other would be equally remarkable. In either case further work on the subject is likely to bring important results in the next few years.

So far I have been discussing only the quasars' radial distribution, which historically is the way their study developed. Now that the Cerro Tololo quasar samples have been completely investigated, the data base at large red shifts is substantial enough for a look at the quasars' three-dimensional distribution. By combining the red shifts with the quasars' positions in the sky one can construct a three-dimensional picture of the quasars in space. This is the only available information on the structure of the universe 13 to 15 billion years ago. The three-dimensional distribution can also yield valuable clues to the nature of the quasars themselves.

If one looks at the distribution of galaxies in photographs showing large areas of the sky, it is apparent that galaxies are anything but uniformly dispersed. They come in pairs, in small groups, in larger groups and in great clusters. Often there are blank regions with few galaxies or none. The study of galactic distribution is crucial to theories of how galaxies form. The current evidence indicates that in the very early stages of the universe matter was evenly distributed as a gas. How, then, did the condensations arise from which stars, galaxies and clusters of galaxies later developed in the course of the expansion of the universe?

Once a condensation has appeared with enough self-gravity to resist the expansion of the universe, one can imagine various ways it might collapse under its own weight, so to speak, to form a star or even an entire galaxy of stars. Most calculations show that once a body of gas begins to collapse the infall of matter to the center is swift and leads to a rapid buildup of density. If the infalling gas is of galactic magnitude, it is easy to imagine that a runaway collapse of matter might create a quasar in the center of the galaxy as part of the galaxy's evolution. This is one possible connection between quasars and the formation of galaxies. Another possibility is that quasars in their own right are powerful enough to influence formation processes on a scale considerably larger than that of a single galaxy. In that case the distribution of quasars might be different from the distribution of galaxies, which we can observe only at a time much later in the expansion of the universe.

To gain a first impression of how the quasars are distributed let us look at the Cerro Tololo data in different pictorial formats. The area of the sky covered in the Cerro Tololo surveys covers a band lying between 37.5 and 42.5 degrees south celestial latitude that extends a fourth of the way around the celestial sphere [see bottom illustration on this page]. The band was recorded in a sequence of about 15 slightly overlapping plates made with the Curtis Schmidt telescope, each plate covering a field five degrees on a side. (The two stars at the end of the Big Dipper that point toward the North Star are separated by about the same angular distance.) The large

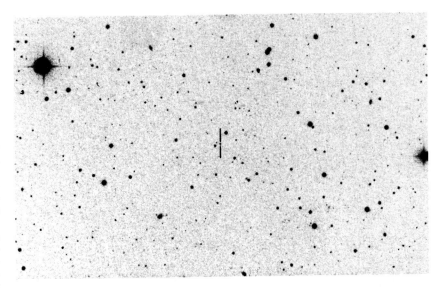

MOST DISTANT OBJECT YET DISCOVERED, on the basis of its red shift of 3.53, is quasar OQ 172, which appears to be an undistinguished star in this photograph made with the 1.2-meter Schmidt telescope on Palomar Mountain. OQ 172 was discovered in 1973 at Lick Observatory by E. Joseph Wampler, Lloyd B. Robinson, Jack Baldwin and E. Margaret Burbidge.

four-meter reflector made deep exposures in seven small regions, each region about one degree square, within the broad band photographed with the Curtis Schmidt. In effect the four-meter telescope sampled long, thin tubes in space.

The Curtis Schmidt survey revealed 88 quasars with red shifts of 1.8 or more. In a much smaller area of sky the four-meter survey identified 53 quasars with comparable red shifts, including six also found in the Schmidt survey. (A handful of the brightest quasars identified in both searches had been discovered earlier by other astronomers.) The Curtis Schmidt survey disclosed one large-red-shift quasar for approximately every four square degrees of sky; the four-meter survey revealed roughly 40 times as many quasars per unit area searched, or about 10 large-red-shift quasars per square degree.

The reason for the higher discovery rate with the four-meter system, of course, is that it could detect considerably fainter objects. The four-meter survey probed to a limiting magnitude of 21, thereby revealing objects about 1.5 magnitudes, or four times, fainter than those that could be recorded by the small Schmidt telescope. Although the

quasars found in the four-meter survey are generally fainter than those found in the Schmidt survey, the fainter objects do not in general exhibit larger red shifts. Most of the 135 quasars in the combined surveys exhibit red shifts near 2. This was expected both because the search method favors the discovery of such quasars and because quasars of that red shift are evidently the most plentiful. The 135 quasars examined in these Cerro Tololo surveys are about a fourth of all the quasars discovered so far with a red shift of more than 1.8.

To get a feeling for the actual distribution of quasars in space I have plotted the 88 Schmidt-survey quasars on a three-dimensional pie-shaped solid [see top illustration on next page]. The front face of the solid is an arc that locates the 88 quasars according to declination, or latitude, and the right ascension (the astronomer's equivalent of longitude, which divides the celestial sphere into 24 slices, each slice 15 degrees wide). The red shift of each quasar is plotted radially on the top of the solid on a scale that runs from a red shift of 1.8 at the near edge to one of 3.5 at the far edge.

Such a plot discloses that in a region

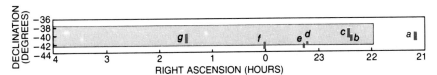

QUASAR SEARCHES AT CERRO TOLOLO were conducted away from the plane of the Milky Way to avoid the obscuring interstellar dust that would dim the light from distant galaxies and quasars. The gray band shows the region of 340 square degrees searched by Smith working with the 60-centimeter Curtis Schmidt telescope equipped with a prism across the aperture. The seven small zones in color, covering a total of 5.1 square degrees, were sampled by Smith and Arthur A. Hoag with the four-meter telescope. The four-meter telescope was able to record objects some four times fainter than any detected by the Curtis Schmidt telescope.

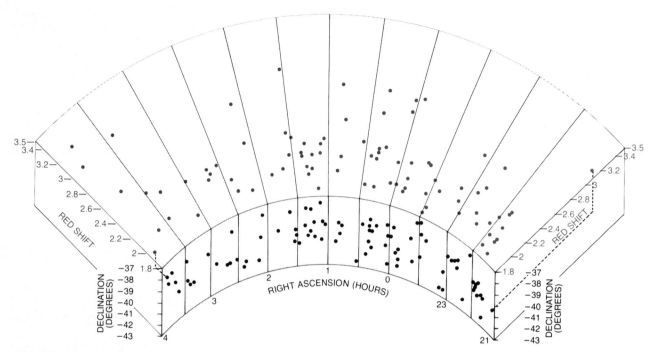

CURTIS SCHMIDT SURVEY located 88 quasars with a red shift larger than 1.8. Since red shifts are correlated with distance, the 88 objects can be plotted as if they were distributed in a three-dimensional volume of space. The black dots on the front face of the diagram show the location of the quasars in the sky. The colored dots on the top surface show the red shift corresponding to each quasar. The plotting scheme is indicated by the broken lines connecting the quasar farthest to the right with its red shift at 3.16 and the quasar farthest to the left with its red shift at 1.96. Other connecting lines have been omitted for the purpose of clarity. The vertical dimension of the surveyed region has been expanded by a factor of two with respect to the horizontal dimension to enhance the separation between quasars.

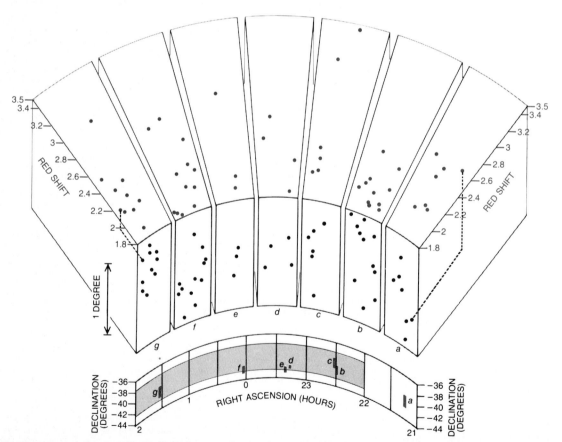

FOUR-METER-TELESCOPE SURVEY located 53 quasars with a red shift larger than 1.8, including six quasars that also turned up in the Curtis Schmidt survey. Here the seven small zones searched with the large telescope are arbitrarily expanded into wedges of uniform volume. The black dots on the front face of the wedges, however, preserve the proper location of each quasar within its own sampling area. Quite a few pairs and small groups of quasars with similar red shifts are evident. In Zone f, for example, four quasars at a distance of about 12 billion light-years are grouped near red shift 1.84. All fall within a volume no more than 200 million light-years across.

between two and four hours right ascension quasars seem to be sparser than they are elsewhere. If the scarcity reflects an actual nonuniformity of quasars in that part of space, it would be startling. Therefore my colleagues and I prefer to think provisionally that the apparent scarcity is due to a selection effect in the survey process. If one looks at the small-scale distribution of quasars, one can see instances where pairs and small groups of quasars lie fairly close together. The most prominent clumping shows up clearly when the quasars in the four-meter survey are plotted in a similar three-dimensional way [see bottom illustration on opposite page]. A group of four quasars with red shifts between 1.83 and 1.86 are clustered in a region no more than 200 million light-years across near zero hours right ascension.

Such groups are fascinating because they have about the same size as the superclusters of galaxies spotted here and there in nearby space. Superclusters are regions some 300 million light-years in diameter that harbor several clusters of galaxies. They are the largest known structures in the universe. So far superclusters have not been observed at red shifts of as much as 2. Could the group of quasars with red shifts around 1.84 mask the location of a remote supercluster? Perhaps the Space Telescope will be able to tell us.

Important as visual examinations of the data are, the question of how quasars are distributed ultimately calls for statistical analysis. Is there an underlying pattern to the distribution? Are the observed groupings merely chance fluctuations? After all, if rice grains are scattered on the floor, some will fall closer together than the average distance between the grains. In recent years powerful statistical techniques have been brought to bear on galaxy surveys to investigate just such questions. With suitable modification for the large red shifts and the selection effects of the survey method such techniques can also be applied to the quasars.

The prime objective is to discover whether the quasars show any departure from a uniform, random distribution. If they do show such a departure, one would like to know whether quasars cluster and what form the clustering takes. Alternatively, one may find that quasars exhibit anticlustering and lie farther apart than one would expect. That could happen if the presence of a quasar at a given place were to inhibit the formation of other quasars nearby. The mathematical formulation of such possibilities and the tests to distinguish them need not concern us here, but the main possibilities lend themselves to graphical presentation in two dimensions. Strong clustering and strong anticlustering are recognizable at a glance. When the clustering or anticlustering is weak, however, the distribution is hard to distinguish from one that is actually uniform and random [see illustration on page 106].

I can summarize the results obtained so far from the Cerro Tololo surveys by saying that the statistical analysis is consistent with the quasars' being distributed uniformly at random. The pairs and groups picked out by eye are evidently nothing more than chance fluctuations. For the 30-odd quasars with red shifts between 1.8 and 2.2 in the four-meter sample the mean nearest-neighbor dis-

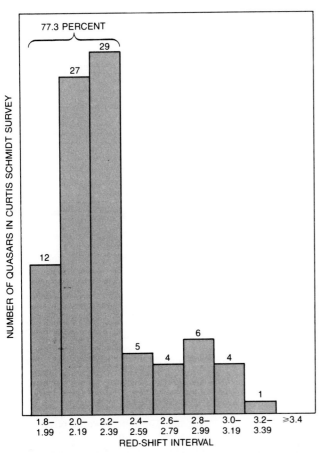

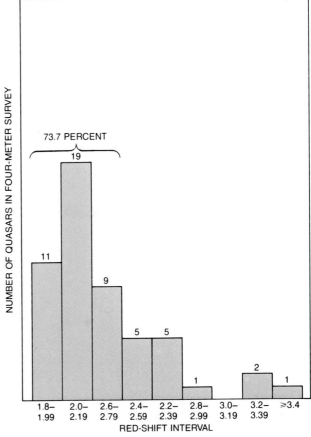

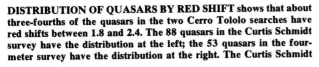

DISTRIBUTION OF QUASARS BY RED SHIFT shows that about three-fourths of the quasars in the two Cerro Tololo searches have red shifts between 1.8 and 2.4. The 88 quasars in the Curtis Schmidt survey have the distribution at the left; the 53 quasars in the four-meter survey have the distribution at the right. The Curtis Schmidt survey disclosed one quasar for every four square degrees of sky. The four-meter-telescope survey, capable of detecting fainter objects, uncovered on the average slightly more than 10 quasars per square degree searched. Although four-meter quasars were generally less luminous than the Curtis Schmidt quasars, they were no more distant.

tance is about 400 million light-years, which is well within the expected range for a random distribution. For the quasars with larger red shifts, where mean separations are measured in billions of light-years, a similar conclusion can be drawn. These findings are evidence in favor of an assumption that is generally held but difficult to confirm, namely that on a large scale the universe is ho-

mogeneous. This assumption is a crucial requirement for current models of the universe.

Reassuring as such results are, they mark only the first step in a continuing inquiry that is likely to progress rapidly as more quasars are found. If the observed clustering of galaxies is extrapolated to the epoch and scale covered in the quasar survey, one can show that

galaxy clustering should not be detectable in the existing samples. It will be interesting to see if the Space Telescope can improve the sensitivity of the distribution tests to the point where galaxy clustering can be detected. The expectation is that galaxy clusters exist at a red shift of 2, but that they stand out less prominently above the background of denser matter in that epoch than do clusters with smaller red shifts observed at a later epoch in a much expanded, and hence less dense, universe. At the same time the apparent pairs and groups of quasars in the four-meter survey cannot be ignored. After all, one theory of galaxy formation postulates that the galaxies originated with random density fluctuations in the early universe. The quasar groupings may represent similar regions of enhanced density.

As a result of our Cerro Tololo investigations one can now try to describe what the universe looked like 12 to 15 billion years ago. Suppose our own galaxy, the Milky Way, were within the group of four quasars that to our present-day instruments exhibit a red shift of about 1.85. Assuming that human life could have evolved that early in the age of the universe, what would we see at night? First of all, the Milky Way itself would be much brighter than it is today because it would have a more prominent population of hot, young stars. The four quasars in the group would be clearly visible to the unaided eye as bright stars; in fact, quasars outside the group, and as far away as 300 million light-years, could also be seen without a telescope. Obviously quasars would have been discovered and recognized as strange phenomena early in the history of astronomy. Whether or not they are the luminous core of galaxies would have been ascertained about as soon as there were telescopes.

It is well to remember that, plausible as our current picture of quasars may be, there is some chance that it is entirely wrong and a good chance that it is wrong in important particulars. Some astronomers doubt that quasars are as distant as their red shifts indicate. Others question the reality of the high density of quasars at large red shifts and propose that quasars were either brighter then or that their intensity has been enhanced as a result of their radiation's passing close to intervening galaxies that act as gravitational lenses. Rarely, if ever, is a large body of data collected and suddenly explained at a stroke by an inspired theory. So it is likely to be with quasars. The subject would be the poorer without the current hypotheses, presumptuous as they may be. In any case it is hard to imagine the prevailing concept of quasars being overturned without their becoming even more amazing than they already are.

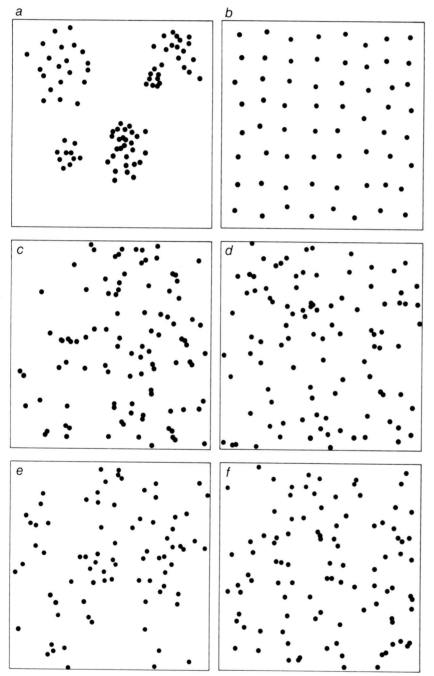

PATTERN OF QUASAR DISTRIBUTION is being sought because it might lead to a fuller understanding of galaxy formation in the early universe. The six patterns appearing in this illustration are two-dimensional analogues of various distribution possibilities. If quasars are strongly clustered (*a*) or strongly anticlustered (*b*), the fact should be apparent. If quasars exhibit only weak clustering (*c*) or weak anticlustering (*d*), extensive statistical tests may be needed to demonstrate that the distributions are not uniform and random. Two computer simulations of uniform, random distributions (*e*, *f*) show that the eye is not a reliable judge of randomness.

THE AUTHORS

BART J. BOK ("The Milky Way Galaxy") was born and educated in the Netherlands. He received his Ph.D. at the University of Groningen. He went to Harvard University in 1929 as Robert Wheeler Willson Fellow in astronomy, and stayed on to become Robert Wheeler Willson Professor of Astronomy and associate director of the Harvard Observatory. In 1957 he left Harvard to become professor of astronomy at the Australian National University and director of the Mount Stromlo Observatory. In 1966 he was appointed head of the department of astronomy and director of the Steward Observatory of the University of Arizona. Shortly before his death in 1983, Bok completed work on the fifth edition of his classic text *The Milky Way;* the first edition, coauthored with his late wife Priscilla F. Bok, was published in 1941.

PAUL W. HODGE ("The Andromeda Galaxy") is professor of astronomy at the University of Washington. His degrees are from Yale University (B.S., 1956) and Harvard University (Ph.D., 1960). Before he joined the faculty at Washington in 1965 he taught and did research at a number of institutions, including Harvard, the Hale Observatories, the California Institute of Technology, the University of California at Berkeley and the Smithsonian Astrophysical Observatory. He is the author of 10 books and more than 200 articles and scientific papers, including "Dwarf Galaxies," which appeared in the May, 1964 issue of SCIENTIFIC AMERICAN. He was awarded the Beckwith Prize in astronomy by Yale in 1956 and the Bart J. Bok Prize in astronomy by Harvard in 1962. In 1978–79 and in 1984–85 he served as chairman of the astronomy section of the American Association for the Advancement of Science. He became editor of *The Astronomical Journal* in 1984. Hodge's major fields of investigation at present include the evolution of stars and galaxies.

VERA C. RUBIN ("Dark Matter in Spiral Galaxies") writes: "For the past 18 years I have been a staff member of the Department of Terrestrial Magnetism of the Carnegie Institution of Washington, D.C., where I work on the dynamics of galaxies. I am also adjunct staff member of the Mount Wilson and Las Campanas Observatories. I obtained degrees in astronomy from Vassar College (a B.A. in 1948), Cornell University (an M.A. in 1951) and Georgetown University (a Ph.D. in 1954). For my M.A. thesis I searched for a large-scale systemic motion of all the galaxies whose radial velocities were then known. For my Ph.D. thesis (under George Gamow) I studied the spatial distribution of galaxies. Both subjects, which are closely related to the work I currently do, were not widely studied again until the 1970's, when more data and large computers became available. My life is busy with travel to observatories, particularly Kitt Peak near Tucson and Cerro Tololo and Las Campanas in Chile."

STEPHEN E. STROM and KAREN M. STROM ("The Evolution of Disk Galaxies") are respectively professor of astronomy and research fellow at the University of Massachusetts at Amherst. Their research activities are divided between studying galaxy evolution and investigating the early evolution of stars. By combining these studies, they hope to achieve a deeper understanding of the star formation process. Stephen Strom received his undergraduate and graduate degrees from Harvard University; Karen Strom is a graduate of Radcliffe College. They have collaborated in preparing over 100 astronomical papers and in raising four children. Besides their mutual passion for astronomy they share an avid interest in photography and they have exhibited their work at galleries throughout the country.

ALAR TOOMRE and JURI TOOMRE ("Violent Tides between Galaxies") are brothers who were born in Estonia and came to the U.S. as boys with their parents. Alar Toomre is professor of applied mathematics at the Massachusetts Institute of Technology; Juri Toomre is professor of astro-geophysics at the University of Colorado at Boulder and a member of the Joint Institute for Laboratory Astrophysics. The brothers did their undergraduate work at M.I.T. and received their Ph.D.'s in England at the University of Manchester and the University of Cambridge respectively. Alar Toomre writes that he likes to "travel and read history, play tennis and bridge and chess and help to raise three children." Juri Toomre writes that his work centers on the "astro" part of his title and that the "geo" part "concerns predicting the structure of the ocean's thermocline." He adds: "I have enjoyed sailing, soaring and the opera; now in Boulder there are the pleasures of living in a high mountain house, hiking and cross-country skiing with the family and a fondness for cooking and Afghan hounds."

JACK O. BURNS and R. MARCUS PRICE ("Centaurus A: the Nearest Active Galaxy") are respectively assistant professor of astronomy at the University of New Mexico and professor of physics and astronomy and chairman of the department at the same institution. Burns was graduated from the University of Massachusetts with a B. S. in 1974; he went on to get his Ph.D. in astronomy from Indiana University. After two years of postgraduate work at the National Radio Astronomy Observatory he moved to New Mexico in 1980. In addition to the subject of the current article his interests include galactic superclusters. Price got a B.S. from Colorado State University before going to Australia to continue his education; his Ph.D. in astronomy was awarded by the Australian National University in 1966. He returned to the U.S. to join the faculty at the Massachusetts Institute of Technology. After eight years he left M.I.T. to become the first radio-spectrum manager at the National Science Foundation. At the NSF he also served as head of the astronomy research section. He went to New Mexico in 1979. He was one of the discoverers of the Faraday effect, whereby the plane of polarization of radio waves from sources outside our galaxy is rotated as the waves pass through magnetic fields in interstellar space.

ROGER D. BLANDFORD, MITCHELL C. BEGELMAN and MARTIN J. REES ("Cosmic Jets") are astrophysicists who share an interest in radio-wave sources that lie outside our galaxy. Blandford, born in England, obtained his B.A. (1970), his M.A. and his Ph.D. (both in 1974) from the University of Cambridge. In 1974 and 1975 he was a member of the Institute for Advanced Study in Princeton. In 1976 Blandford moved to the California Institute of Technology, where he is professor of theoretical physics. Among his research interests are the physics of extragalactic radio sources, cosmic rays and neutron stars. Begelman is

assistant professor of astronomy at the University of Colorado and a member of the Joint Institute for Laboratory Astrophysics. He received both his B.A. and his M.A. at Harvard University in 1974. His Ph.D. was granted in 1978 by the University of Cambridge, where he has returned for several periods since. He went to Berkeley in 1979, and Rees, like Blandford a native of England, is Plumian Professor of Astronomy at Cambridge and director of the Institute of Astronomy there. He got his B.A. (1963) and Ph.D. (1967) from Cambridge. He has held visiting appointments at Cal Tech, the Institute for Advanced Study and Harvard, and he was a member of the faculty of Sussex University before taking up his present post in 1973.

STEPHEN A. GREGORY and LAIRD A. THOMPSON ("Superclusters and Voids in the Distribution of Galaxies") are astronomers with a particular interest in the structure and evolution of galaxy superclusters. Gergory's bachelor's degree is from the University of Illinois at Urbana-Champaign; his doctorate in astronomy is from the University of Arizona. From 1973 to 1977 he was on the faculty of the State University of New York at Oswego. In 1977 he went to Bowling Green State University as assistant professor of physics. Thompson received his undergraduate training in astronomy and physics at the University of California at Los Angeles. His Ph.D. in astronomy was awarded by the University of Arizona in 1974, the same year as Gregory's. After two years at the University of Nebraska he moved to the University of Hawaii at Manoa, where he is currently a member of the staff of the Institute for Astronomy. His research interests include (in addition to superclusters) supernovas, high-resolution imaging and the color photography of celestial phenomena.

PATRICK S. OSMER ("Quasars as Probes of the Distant and Early Universe") is director of the Cerro Tololo Inter-American Observatory in Chile. He majored in astronomy at the Case Institute of Technology, obtaining his bachelor's degree in 1965. His Ph.D., also in astronomy, was awarded by the California Institute of Technology in 1970. He began at Cerro Tololo in 1969 as a research associate; he was appointed director last year. The observatory is the site of a four-meter telescope; installed in the 1970's, it is the largest in the Southern Hemisphere. His research interests include the atmospheres of very luminous stars, the Magellanic clouds and stellar sources of X rays. Quasars have been Osmer's major interest since 1974.

BIBLIOGRAPHIES

1. The Milky Way Galaxy

STAR FORMATION. Edited by T. de Jong and A. Maeder. D. Reidel Publishing Company, 1977.

THE LARGE-SCALE CHARACTERISTICS OF THE GALAXY. Edited by W. B. Burton. D. Reidel Publishing Company, 1979.

INTERSTELLAR MOLECULES. Edited by B. H. Andrew. D. Reidel Publishing Company, 1980.

STAR CLUSTERS. Edited by James E. Hesser. D. Reidel Publishing Company, 1980.

THE MILKY WAY. Bart J. Bok and Priscilla F. Bok. Harvard University Press, 1981.

2. The Andromeda Galaxy

SPIRAL STRUCTURE IN M31. Halton Arp in *The Astrophysical Journal*, Vol. 139, No. 4, pages 1045–1057; May 15, 1964.

GALAXIES. Harlow Shapley. Revised by Paul W. Hodge. Harvard University Press, 1972.

A RADIO CONTINUUM SURVEY OF M31 AT 2695 MHz, I: OBSERVATIONS; COMPARISON OF RADIO CONTINUUM DATA. Elly M. Berkhuijsen and R. Wielebinski in *Astronomy and Astrophysics*, Vol. 34, No. 2, pages 173–179; August, 1974.

THE OPEN STAR CLUSTERS OF M31 AND ITS SPIRAL STRUCTURE. Paul W. Hodge in *The Astronomical Journal*, Vol. 84, No. 6, pages 744–751; June, 1979.

STUDIES OF LUMINOUS STARS IN NEARBY GALAXIES, IV: BAADE'S FIELD IV IN M31. Roberta M. Humphreys in *The Astrophysical Journal*, Vol. 234, No. 3, Part 1, pages 854–860; December 15, 1979.

3. Dark Matter in Spiral Galaxies

DYNAMIC EVIDENCE ON MASSIVE CORONAS OF GALAXIES. Jaan Einasto, Ants Kaasik and Enn Saar in *Nature*, Vol. 250, No. 5464, pages 309–310; July 26, 1974.

MASSES AND MASS-TO-LIGHT RATIOS OF GALAXIES. S. M. Faber and J. S. Galagher in *Annual Review of Astronomy and Astrophysics*, Vol. 17, pages 135–187; 1979.

THE LARGE-SCALE STRUCTURE OF THE UNIVERSE. J. P. E. Peebles. Princeton University Press, 1980.

4. The Evolution of Disk Galaxies

SPIRAL STRUCTURE, DUST CLOUDS, AND STAR FORMATION. Frank H. Shu in *American Scientist*, Vol. 61, pages 524–536; 1973.

GALAXIES AND THE UNIVERSE, VOL. 9: STARS AND STELLAR SYSTEMS. Edited by Allan Sandage, Mary Sandage and Jerome Kristian. The University of Chicago Press, 1975.

THE FORMATION OF GALAXIES. Richard B. Larson in *Galaxies: Sixth Advanced Course of the Swiss Society of Astronomy and Astrophysics*, edited by L. Martinet and M. Mayor. Geneva Observatory, 1976.

CLUSTERS OF GALAXIES. Neta A. Bahcall in *Annual Review of Astronomy and Astrophysics*, Vol. 15, pages 505–540; 1977.

THE ORIGIN OF GALAXIES. Richard B. Larson in *American Scientist*, Vol. 65, pages 188–196; 1977.

THEORIES OF SPIRAL STRUCTURE. Alar Toomre in *Annual Review of Astronomy and Astrophysics*, Vol. 15, pages 437–478; 1977.

5. Violent Tides between Galaxies

MULTIPLE GALAXIES. F. Zwicky in *Ergebnisse der exakten Naturwissenschaften*, Vol. 29, pages 344–385; 1956.

GRAVITATIONSEFFEKTE BEI DER BEGEGNUNG ZWEIER GALAXIEN. J. Pfleiderer in *Zeitschrift für Astrophysik*, Vol. 58, No. 1, pages 12–22; August, 1963.

ATLAS OF PECULIAR GALAXIES. Halton Arp. California Institute of Technology, 1966.

GALACTIC BRIDGES AND TAILS. Alar Toomre and Juri Toomre in *The Astrophysical Journal*, Vol. 178, pages 623–666; December 15, 1972.

TIDAL INTERACTION OF GALAXIES. T. M. Eneev, N. N. Kozlov and R. A. Sunyaev in *Astronomy and Astrophysics*, Vol. 22, No. 1, pages 41–60; January, 1973.

6. Centaurus A: the Nearest Active Galaxy

THE INVISIBLE UNIVERSE. Gerritt L. Verschuur. Springer-Verlag, 1974.

ACTIVE GALACTIC NUCLEI. Edited by C. Hazard and S. Mitton. Cambridge University Press, 1979.

THE X-RAY STRUCTURE OF CENTAURUS A. E.D. Feigelson, E. J. Schreier, J. P. Delvaille, R. Giacconi, J. E. Grindlay and A. P. Lightman in *The Astrophysical Journal*, Vol. 251, No. 1, Part 1, pages 31–51; December 1, 1981.

EXTRAGALACTIC RADIO SOURCES. Edited by David S. Heeschen and Campbell M. Wade. D. Reidel Publishing Company, 1982.

THE X-RAY JETS OF CENTAURUS A AND M87. Eric D. Feigelson and Ethan J. Schreier in *Sky and Telescope*, Vol. 65, No. 1, pages 6–12; January, 1983.

CENTAURUS A. K. Ebneter and B. Balick in *Publications of the Astronomical Society of the Pacific*, Vol. 95, pages 675–690; 1983.

7. Cosmic Jets

THE STRUCTURE OF EXTENDED EXTRAGALACTIC RADIO SOURCES. George Miley in *Annual Reviews of Astronomy and Astrophysics*, Vol. 18, pages 165–218; 1980.

THE BIZARRE SPECTRUM OF SS 433. Bruce Margon in *Scientific American*, Vol. 243, No. 4, pages 44–55; October, 1980.

RELATIVISTIC JET PRODUCTION AND PROPAGATION IN ACTIVE GALAXIES. M. J. Rees, M. C. Begelman and R. D. Blandford in *Annals of the New York Academy of Sciences*, Vol. 375, pages 254–286; 1981.

8. Superclusters and Voids in the Distribution of Galaxies

THE COMA/A1367 SUPERCLUSTER AND ITS ENVIRONS. Stephen A. Gregory and Laird A. Thompson in *The Astrophysical Journal*, Vol. 222, No. 3, pages 784–799; June 15, 1978.

THE COSMIC TAPESTRY. Guido Chincarini and Herbert J. Rood in *Sky and Telescope*, Vol. 59, No. 5, pages 364–371; May, 1980.

THE PERSEUS SUPERCLUSTER. Stephen A. Gregory, Laird A. Thompson and William G. Tifft in *The Astrophysical Journal*, Vol. 243, No. 2, Part 1, pages 411–426; January 15, 1981.

9. Quasars as Probes of the Distant and Early Universe

QUASI-STELLAR OBJECTS. Geoffrey Burbidge and Margaret Burbidge. W. H. Freeman and Company, 1967.

THE CLUSTERING OF GALAXIES. Edward J. Groth, P. James E. Peebles, Michael Seldner and Raymond M. Soneira in *Scientific American*, Vol. 237, No. 5, pages 76–98; November, 1977.

QUASARS: OBSERVED PROPERTIES OF OPTICALLY SELECTED OBJECTS AT LARGE REDSHIFTS. Malcolm G. Smith in *Vistas in Astronomy*, Vol. 22, pages 321–362; 1978.

OBJECTS OF HIGH REDSHIFT. George O. Abell and P. J. E. Peebles. D. Reidel Publishing Company, 1980.

INDEX